T0073174

Communication Between Honeybees

Jürgen Tautz

Communication Between Honeybees

More than Just a Dance in the Dark

Translated by David C. Sandeman

 Springer

Jürgen Tautz
University of Würzburg
Waldbrunn, Bayern, Germany

ISBN 978-3-030-99483-9   ISBN 978-3-030-99484-6   (eBook)
https://doi.org/10.1007/978-3-030-99484-6

This Springer imprint is published by the registered company Springer Nature Switzerland AG
The registered company address is: Gewerbestrasse 11, 6330 Cham, Switzerland

Preface

What It's About

An old joke tells of a man in a dark alley searching for his house keys beneath a street lamp, helped by a friendly passer-by. After an unsuccessful search, the passer-by asked the unfortunate man where he had lost his keys. The man replied that he had lost them somewhere on the way but here there was finally enough light to look for them.

In Science too, from time to time and in the absence of appropriate research methods, early explanations and solutions are not necessarily sought where true answers are hidden. If early incomplete explanations are attractive, they can persist for a very long time.

Bee research over the past hundred years has provided us with a deep insight into the homes of honeybees. We have learned a great deal about how they live and how they survive in their world. The study of honeybees has had a strong influence on the development of modern behavioural research and concepts. Communication biology in particular has significantly profited from an increased interest, and the so-called dance language of the honeybee is still accepted. In well over one thousand scientific publications about the dance language, insights have been won and concepts developed that are incorporated in countless school and learning texts and Internet contributions.

The function and importance of the waggle dance are overvalued. If a bee enters a hive and dances, so the classic story goes, she communicates with her colleagues and passes on information that will lead them directly to a food source. The second half of the story, namely the communication between bees outside the hive, continues to remain practically ignored. It plays a minor role in concepts and models of the complex communicative behaviour of the honeybee and constitutes a blind spot in bee research. Honeybees are not social insects only within their hives—if one takes this into account, it would appear that the emphasis in bee communication research has so far been particularly one-sided.

The renowned insect researcher Edward O Wilson wrote in 1971.

"Furthermore, the waggle dance had become something of a sacred cow and it needed a critical examination by an independent group of investigators." (287, p. 267)

And further:

"Also, there is a scarcity of measurements of the amount of information added to the waggle dance by additional cues, in particular the assembly pheromones of the Nasanov glands released in the vicinity of the new finds and the sight of flying workers." (287, p. 268)

Here, it concerns the communication between foraging bees in the field.

With the award of a Nobel Prize to Karl von Frisch in 1973, this "sacred cow" became firmly established and Wilson's criticism went unheard. The attention he drew at the time to the missing second half of research on the recruitment of foragers is valid to this day.

In the literature, one can indeed find the consideration that in addition to the dance bees do obtain help from other sources to guide them to a goal. However, such comments remain without consequence because they are neither included in the concept of how recruitment of foragers proceeds, nor followed up experimentally.

This book analyses the state of our knowledge from published studies of the bee dance, orders essential elements from these into a conceptual overview, and develops a program for necessary research to eventually complete the picture of one of the most remarkable behavioural achievements in the animal kingdom.

This book does NOT take into account how important, in general, the advertisement for a food source is for a bee colony, when information from the dance is ignored, and how this can change with the circumstances.

Bees that follow a dance have many options for the associated flight—nevertheless, based on information from the dance alone, they would not find the desired goal.

This book examines the mechanisms about how a goal for which a dancer advertises is nevertheless found. It illuminates the core of the dance language. If this is critically exposed, many thoughts and publications about the dance really being a language are irrelevant. The book focuses, more modestly, merely on how new recruits get to a food source for which a forager advertises in her dance.

Waldbrunn, Germany   Jürgen Tautz

Acknowledgements

Many have contributed to the development of the difficult and complex contents of this book.

I thank David C. Sandeman for his patient and constructively critical support over many years, for his confidence in the realisation of this book, and for his excellent translation into English.

To Rodrigo De Marco, Benjamin Rutschmann, and Stephen Buchmann, my thanks for the insights into and agreeing to the use of results of not yet published projects. I thank Werner Nachtigall, joint founder of the Gesellschaft für Technische Biologie und Bionik, for information about kinematics and dynamics of bee flight that could be relevant for the function of buzzing flights. I thank Michael Boppre for advice on the literature and for joint speculation on the possibility of a graded release of Nasanov gland pheromone. I thank Martin Wikelski for discussions about distant goal orientation in birds. My thanks go to Silke Arndt for her creativity and patience in developing the graphics. I thank the publishing teams at Knesebeck and at Springer for an excellent collaboration.

Without the support of my wife, Rosemarie Müller-Tautz, also in the difficult phases of agonising doubt about my own conclusions, this book would not have been written.

Contents

About the Author

Prof. Dr. Jürgen Tautz is a bee expert, sociobiologist, behavioural researcher, and Professor in retirement at the Biozentrum, University of Würzburg. He is the Chair of the Bee Research Würzburg e.V. and the Head of the interdisciplinary project Honey Bee Online Studies (HOBOS) and the follow-up project we4bee. He is a bestseller author and frequently cited for his successful communication of science to a broad public.

The Struggle for Insight

1

Those who have treated the sciences have been either empirics or dogmatical. The former like ants only heap up and use their store, and the latter like spiders spin out their webs. The bee, a mean between both, extracts matter from flowers of the garden and the field, but works and fashions by its own efforts. Sir Francis Bacon (1620)

New data does not always bring change in science, but instead new perspectives—in keeping with the parable of Francis Bacon. A new objective view of known facts can lead to completely new interpretations and understanding.

A wonderful example is the discovery of how to split the atom. Otto Hahn experimented with uranium in his Berlin laboratory and stayed in contact with his colleague, Lisa Meitner, who had fled to Sweden to avoid persecution from the National Socialists, through the exchange of letters about his research. Doubt surfaced frequently in his communications about his difficulty in reconciling his findings with the basic ideas that prevailed at the time. Lisa Meitner thought about the problem, calculated, and concluded that it must be possible to shatter the atomic nucleus. The view beyond the notion of indivisible atoms led to completely new ideas and the discovery of the fissionable atom.

Another well-known example of a change of viewpoint with far-reaching consequences was the recognition of our heliocentric planetary system. The Greek astronomer Ptolemy collected exact data on the movement of planets because very complicated assumptions and explanations were proposed why, at times, they travelled backwards. He had no other alternative possibilities other than to contribute further assumptions, as long as he held to the basic premise that the earth was the centre of the Universe. When Nikolaus Kopernicus examined Ptolemy's

The original version of this chapter was revised for adding the Extra Server Material. The correction to this chapter can be found at https://doi.org/10.1007/978-3-030-99484-6_13.

Supplementary Information The online version contains supplementary material available at https://doi.org/10.1007/978-3-030-99484-6_1.

data from a different viewpoint and considered the revolutionary idea that the earth circled the sun instead, a much simpler explanation suddenly arose for the motion of planets and a new concept of the Universe.

The living world lies between the smallest and largest elements of the physical world that man has explored. Here too, the viewpoint taken when observing data and phenomena can be critical.

Over the last decades, a discovery in the life sciences with significant consequences was the molecular structure of genes. Maurice Wilkins and Rosalind Franklin had obtained X-ray crystallographic information on deoxyribonucleic acid from the cell nucleus of living organisms, but did not view their data from the same point of view as their colleagues, James Watson and Francis Crick. Their new view of available results and a little work with scissors, paper, and glue led to the idea of a double helix as the basic structure of hereditary material. With this, they created the basis of modern molecular genetics and in 1962 were together with New Zealander Maurice Wilkins awarded the Nobel Prize for medicine.

A different viewpoint of the dance language of honey bees is also possible and this is taken here. A new ordering of old data and cognisance of gaps in our knowledge accompany this book throughout as a constant reminder during the review of observations, experiments, and interpretations that have decisively led to the present picture of bee communication.

Why Half-Truths Are Practical

The bee dance is a form of communication between honeybees. A classic formulation of this probably best-known form of communication in the animal world found in every text about the bee dance is as follows: "In their dance, honeybees provide the direction and the distance to a goal." Why is this strict statement a half-truth? In fact, bees do not communicate information about the position of the food source in their dance. Correct is that the dance provides a rough indication of the direction of and the distance to a geographical area.

Formally and correctly expressed, although sounding somewhat affected, is that through the dance the bees communicate a suggestion that outside the hive a new recruit will receive goal orienting signals and cues to a food source that can be found at a particular angle and distance in relation to the hive.

In the language of communication science, the bee dance takes the first step to reduce the uncertainty about where in the field the dance follower will have a high possibility to meet connecting signals and cues. These stimuli offer the new recruit the second step and the essential information that lead to the goal. This dry description of the communication biological meaning of the bee dance is the theme for the design of this book. A description of the complicated dance information is as follows: "Through the dance information honeybees reduce the uncertainty about where in the field recruits will meet with orienting signals and cues that will lead them to the goal." This does not sound as elegant as the classical explanation of the bee dance but it is correct.

The classical formulation is simple, graphic, and easy to understand. That makes it attractive, but it is a conceptual model of reality and carries the inherent danger that the model, in our minds, becomes reality. Is it hair-splitting to be so exact in formulating what the bee dance achieves? What are the consequences of not being exact? Where did the half-truth statement come from? Why are they so persistent? This book will provide the answers.

So How Does a Forager Find the Goal?

There are abbreviated statements about factual issues that no one would think about taking literally. If one took the statement that the world is 4.6 million years old, one could calculate the exact date and day of the week of its coming into being. Clearly absurd and obvious to all is that this estimate of its age only establishes an approximation.

In what way is the statement "Honeybees provide the direction and the distance to the food source in their dances" (in the dance language) any different? Is the statement taken seriously?

Implicit is the idea that dancers not only convey the direction and the distance to the goal but also that the recruited forager can find a previously unknown food source given the information contained in the dance. Research on the dance language of the honeybees and also this book are concerned with exactly this aspect: How does a forager find its way to a goal to which she was recruited by a dancer?

Over the last 2000 years, there have been essentially three suggestions:

1. The recruits are LED to the goal (Aristotle).
2. The recruits are SENT to the goal (Karl von Frisch = dance language).
3. The recruits are ATTRACTED to the goal (Adrian Wenner).

In the following, it will be shown that all three of these ideas are valid but not alternatives. Instead, the sending, leading, and attraction are interwoven links in a complex communicative chain beginning in the hive and ending in the goal. All three together are necessary to bring the recruits to the geographical location of the goal.

The special area of behavioural science concerned with the orientation abilities of animals has developed a number of general concepts and definitions that apply to the honeybee's search for a goal. Research into orientation over long distance, where animals achieve an impressively high performance, presents a particular challenge. These reach a spatial goal they are unable to see at the start nor directly perceive on the first leg of their journey. The goal at the beginning is not seen, nor smelled, heard, or perceived in some other way. There is no natural connection with the destination.

The analysis of typical characteristics of navigation was researched in early studies of homing in birds, several years before the beginning of modern research on communication in honeybees [222, 277].

Maximally three sequential phases are involved in which an animal can find its way to a distant goal. To begin, a direction is chosen and maintained over a certain distance, which then gives way to a search phase, followed by a directed orientation that leads the animal to its goal.

An explanation of how the new bee recruits arrive at a goal, to which they have no sensory connection at the start from the hive, will become ever clearer with the passage of this book and in relation to the model ornithologists introduced very early for bird migration. The concept "In their dance, honeybees provide the direction and the distance to a goal" contains no differentiation into the three phases of distant orientation because, it is supposed, the dance follower can derive the position of the goal from the dance.

In fact, the dance does not achieve this. Instead, its contribution is only the first phase of distant goal orientation, not the second and third phases. The dance sends the recruit on its first phase to a region where the second phase of distant goal orientation begins. There, recruits search for leading signals and cues such as the scent of flowers and communication with experienced foragers (Fig. 1.1).

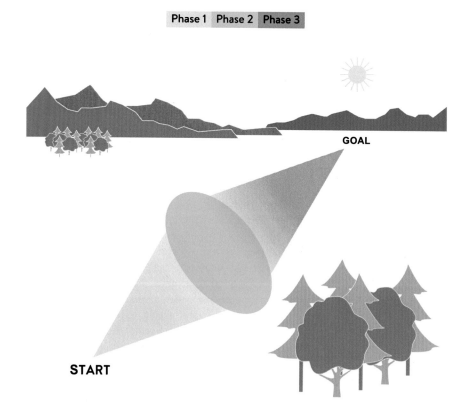

Fig. 1.1 Distant goal orientation leads a bee over three phases from the start to the goal. Phase 1, fly off in a chosen direction (SEND—yellow sector), until Phase 2, when the search area is reached and goal orientation stimuli are available (SEARCH—green sector), transfer to Phase 3 (ATTRACT—orange sector) follows

Initially, it may seem an exaggeration, given the stretches bees cover, to talk of distant goal orientation. This is very different from the bee's point of view. A ten-kilometre distant goal is, roughly calculated, about one million bee body lengths away. Calculating this for the body length of a white stork would mean a distance of a thousand kilometres for the bird, a flight that deserves to be called distant.

How Research Explains the Phenomenon

In addition to the dance language, a second hypothesis has been proposed to explain how a recruit reaches a goal for which a dancer has advertised. This hypothesis (scent of the goal alone attracts bees) also does not differentiate between consecutive orientation phases. It assumes the first phase does not exist and that recruits search for and then simply follow scent trails that lead to the goal.

Dictionaries bring the state of research on recruitment of foragers, slightly modified and always back to the same point, for example in the Cambridge Dictionary (1995):

> There are two main hypotheses to explain how foragers recruit other workers—the waggle dance or dance language theory and the odour plume theory.

In fact, neither of the two hypotheses is correct.

Neither approach resulted in a research program in which all three phases of distant goal orientation (initial stretch, search step, and orientation step) that a new recruit takes to reach the goal receive the same attention and are investigated.

It is interesting to examine the observations, thoughts, and experiments from which the researchers derive their conclusions. Different standpoints are established not only through individual intellectual dead ends, but also through the power of terms used and lack of preparedness to accept and integrate results and data that do not "fit."

To make a theme or area lively and transparent in this book, which is as complex as the research and ideas surrounding the bee dance, the individual researchers should speak for themselves. Experiments and their results are presented and the conclusions researchers derived (or did not derive) are described.

The suspected record number of scientific publications about the bee dance makes a selection of available publications unavoidable. However, care has been taken to represent each standpoint with its central experiments, data, and implications.

It is illuminating, when reading publications that are concerned with the bee dance, to see which follow the central issue of this book and if the aspect of communication between bees in the field takes place are mentioned in the study or even given attention in practice.

It seems an irony in the history of bee research that systematic research of honeybee behaviour began more than a hundred years ago with the discovery of communication between foragers in the field. This discovery was, although not corrected, simply overridden.

As in every science, next to the fantasy and observational skills of researchers, there are available methods appropriate to the state of knowledge and ideas that constitute a phenomenon. See the old joke in the Preface.

To investigate communication between honeybees means the methods must be appropriate for the study of bees both in their nest and also out in the field.

Next to technical observational methods that are continually expanding, the ability of honeybees to learn opens the way for research possibilities found in very few other insect species. In learning experiments, bees allow themselves to be directly "questioned" about the perceptive world they live in. Karl von Frisch was the first to recognise and exploit this.

Thus equipped, bee research worked its way ever deeper into the world of the honeybee. One of the most stimulating fields is the communication biology of these insects. A few animals and no other insects have been the subject of as many behavioural studies as the honeybee.

Honeybees Make No Clear Statements in Their Dances

The particular interest of science in honeybees and the fascination they evoke are reasons why the so-called dance language is more popular than any other animal behaviour. When reading or learning school texts or consulting the Internet about the dance language, one is immediately confronted with the formulation provided at the beginning of this book—it is set in stone.

This would not be hard to accept if the formulation was understood for what it is, namely an oversimplified and strongly reduced model that does not represent biological reality.

Bohr's model of the atom with a central nucleus and electrons that circle in shells around it is a good start into the world of the atom with unquestioned didactic value and therefore popular to this day. For modern atomic physics, this model with its half-truths that do not represent the reality of the atom is merely of historical importance. Will the half-truths of the bee dance also end in that way?

An area of mathematics, fuzzy logic, is concerned with the treatment of unclear statements [139]. The roots of fuzzy logic go back to the Greek philosopher Plato who expressed the thought that between the statements "true" and "false" lay a third possibility, the "half-true." In contrast, his colleague Aristotle was of the opinion that the precision of mathematics and science was only possible if statements were either "true" or "false."

Honeybees with half-true statements in their dance are clearly on Plato's side. Half-truths in the bee dance are not untruths.

How Research on Honeybee Communication Began

2

From Aristotle to Maeterlinck

> According to Bernhard von Chartres we are like dwarfs sitting on the shoulders of giants to see more and further than they could – not thanks to our sharper eyes or small bodies but because the size of the giants elevates us. (Johannes von Salisbury, 1159)

In the last 2000 years of bee research, the apparently perfect collaboration of worker bees, in particular, attracted much attention. At first, bees collecting nectar and pollen were observed. Soon after, however, interest was awakened about events within the bee nests.

Aristotle Discovers That Bees Are Not Independent Individuals

As with so many themes, one can begin with the recorded observations of Aristotle (384–322 B.C.).

The results of his field studies on bees are to be found in his *Historia animalum* written in 400BC:

> On an excursion a bee does not select a specific type of flower but flies from bloom to bloom (wallflower, snowdrops, or violets), and does not contact other bees until she returns to her hive. When she arrive, she shakes herself, with three or four other workers close to her sides. What she brought with her is not easily observed and how she carries out her foraging has not yet been seen because [the workers] spend a long time in the same place among the many leaves. ([5, p. 76])

The available text is not precise. Nevertheless, if the forager shakes herself when she arrives at the hive, then Aristotle was the first to describe the bee dance; if at the flowers, then he was the first to describe mixed groups of experienced bees and recruits. Aristotle is cited in the literature only in relation to the second possibility.

© The Author(s), under exclusive license to Springer Nature Switzerland AG 2022
J. Tautz, *Communication Between Honeybees*,
https://doi.org/10.1007/978-3-030-99484-6_2

It is not clear though whether all authors referred to the original work or relied on a previous translation.

Butler Describes How Flying Bees Detect Odours

Charles Butler (1560–1647) contributed the following observation to the biology of orientation and communication in honeybees:

> But their sense of smell is excellent, by which when they fly aloft in the air, they will quickly perceive anything under them that they like, such as honey, resin or tar, even though it is covered. As soon as the honey-dew is fallen, they immediately change direction, even though the oaks which receive it are far off. [24, p. 20 f.]

Swammerdam Investigates Bees in Their Nests

The Dutch, Jan Swammerdam (1637–1680) was a bee research pioneer. Being one of the first to use a microscope, he described the form and anatomy of bees and confirmed the discovery of the Spaniard, Luis Mendez de Torres [264] that bee colonies have a queen, no king, and she alone lays eggs.

Swammerdam slipped sheets of white paper between the hive combs in order to better observe the behaviour of bees in a busy hive.

Reaumur Invents the Glass Observation Hive

Swammerdam's observational techniques did not allow him to discover the origin of wax. The answer to this puzzle was first published after Rene-Antoine Ferchault de Reaumur (1683–1757) opened a new world to bee research by the introduction of an observation hive with glass windows (Fig. 2.1). Plinius (Gaius Plinius Secundus) though reported bee hives with windows of thin, transparent horn [87] although no illustrations have been handed down.

Reaumur developed a variety of different models of his revolutionary hive. In his 1740 book on honeybees, he writes:

> The invention of the glass or transparent bee hive is new. It was apparently unknown in 1680, in Swammerdam's time [...]. Swammerdam would have without doubt made many more observations on bees but which he could not have seen without such a glass observation hive. [201, pp. 11, 12]

Reaumur himself studied the construction of combs and worker to queen-related behaviour in an attempt to understand how activity in a hive was initiated and controlled. In his book on observation hives, he continued:

Fig. 2.1 One of the first observation hives. The etching of the observation hive in the Medical garden in Amsterdam was published in an Encyclopaedia in 1730

It has to be admitted that a casual viewer of an observation-hive […] would not find much satisfaction. To his regret, the manoeuvre that he would so like to see often takes place at a poorly illuminated site and hidden from his view. In general, the worker bees are all far too busy. Individual bees on which an observer focuses, and would like to watch, are soon hidden by others. Activity in a well populated bee hive is even more extreme and confusing although an order prevails. [201, p. 12]

Spitzner Observes the Dance and Buzzing Flights

A minimal version of the observation hive, consisting of four combs, arranged above and next to one another and enclosed between two glass panes, is necessary for a significant improvement in the observation of bee behaviour and of individual bees over an extended period. Thus equipped, Johann Ernst Spitzner (1731–1805) was the first bee researcher not only to provide relevant information about the recruitment behaviour of honeybees, but also the first to carry out experiments. It is worth reading what he wrote about the first experimental study on communication between honeybees (Fig. 2.2):

> For foraging, only those bees that can be spared from care and warming the brood take part; on cool days only a third of the population but on warm days and when there is a lot to harvest, as much as half the hive. A returning bee that has found a good honey source rapidly informs others who hurry after her when she flies out again. I explored this with the observation hive as follows: I set a small dish of honey in the grass at the end of the garden,

(...) Wenn eine Biene irgendwo gute Honigtracht gefunden hat, so macht sie es den anderen geschwind bekannt, und sie eilen alle dieser zuerst heimgekommenen in vollem Fluge nach, wenn sie wieder auffliegt. Damit geht es so zu, wie ich es in den Glasstöcken erforscht habe: Ich setzte ein kleines Gefäß mit Honig am Ende des Gartens in das Gras, trug nur zwei Bienen aus dem Glasstocke dazu, daß sie sich davon vollsaugen konnten, und gab Acht, wie sich diese bei ihrer Rückkunft in den Stock gebehrden würden. Sie kamen bald zurück, schwungen sich länger als andere mit einem hellsingenden Ton um das Flugloch herum, und als sie zu den anderen eingegangen waren, wälzten sie sich öfters von unten bis oben, und von oben bis wieder herunter auf den an den Tafeln ruhig sitzenden Bienen in beständigen Kreisen herum, reichten auch mancher ihre Zunge zum Ablecken, ehe sie ihr gebrachtes Honig in eine Zelle ausschütteten. Dadurch geriethen alle in Allarm, liefen nach dem Flugloche zu, und folgten dieser, wieder mit einem hellsingenden Ton abgehenden Bienen in vollen Haufen nach.(...)

Aus Spitzer 1788, S. 69

Fig. 2.2 Page 69 from Johann Ernst Spitzner's book *Ausführliche Beschreibung der Korbbienenzucht im sächsischen Churkreise*, Leipzig 1788 [245]

carried two bees from the observation hive to the dish, let them feed, and watched to see how these conducted themselves on their return to the hive. They soon returned and, emitting a high frequency tone (this could only be an example of a buzzing flight, J.T.) hovered in front of the hive entrance longer than usual before joining the others within. There they waltzed up the surface of the combs and down again among the bees sitting quietly in circles around them. Some reached out their tongues to be licked before they deposited the honey they had brought with them into cells. The result was a general alarm in which bees crowded to the hive entrance to fly in a cluster after the again high tone-emitting bees. A few minutes later the small dish was covered with bees, the honey taken up and transported into the hive. This took place on a dull day on which they did not go out into the field, so that other bees (i.e. from other hives, J.T.) were not aware. Nevertheless, several bees from nearby hives were also attracted by the high tone, and followed to the source. [245, pp. 68–70]

Sprengel Discovers That Insects Pollinate Flowers

One can find no further studies of bee behaviour in the field for almost 2000 years after Aristotle. Christian Konrad Sprengel (1750–1816) was the first to publish again and with epochal insight (Fig. 2.3). The realisation that insects pollinate flowering plants and that honeybees play an important role goes back to him.

Aristotle was, as mentioned earlier, aware that bees visited flowers but he gave the reason for this as follows:

> Worker bees take up wax from parts of the plants with their forelegs, brush these off with the middle legs and the middle legs on swellings of the hind legs. They fly off, clearly heavily loaded. ([5, p. 76])

This behaviour, mistakenly taken for wax collection, perfectly describes the collection of pollen. The true source of wax was first appreciated when a way to observe bees in their nests was found.

Fig. 2.3 The cover of Christian Konrad Sprengel's epochal book *Das entdeckte Geheimnis der Natur im Bau und in der Befruchtung der Blumen*, Berlin 1793 [244]

Huber's Revolutionary Observations—Although He Was Blind

Francois Huber (1750–1831) had an ingenious idea. Blind from youth, he designed a hive in which the single combs could be turned like the pages of a book. Following his directions, his wife and his servant, Franz Burnens told him what they saw the bees to be doing. Through the eyes of his helpers, he was able to discover that virgin queens were not fertilised by drones in their own hive but had to leave the hive and that pairing took place in the air. Huber also contributed significant insights into the source of wax plates, the production of which was first directly observed by Pastor Herman Christian Hornborstel from Dörverden in 1720. In addition, Huber thoroughly studied and described the exchange of air in the hive and fanning behaviour of bees (Chapter "Nachforschung über die Art der Erneuerung der Luft in den Stöcken," in [122], p. 175ff.) (Fig. 2.4).

> I decided to fasten a light-weight wind-sensitive structure, for example a strip of paper, feather or wool, at the entrance to the hive. Suspended from a thread on a stick these should show if a measurable airflow was present and of what strength […]. The sensors were scarcely in position before they began to move; they first moved against the entrance and remained there for a moment, then soon after moved back with the same velocity to hang one or two inches from vertical in the air. This inflow and outflow appeared to be related to the number of bees that were fanning; they were less active at times but never stopped entirely. [122, p. 175ff.]

It is surprising that despite the immense number of precise observations in both Huber's books, no single description appears that could relate to the dance. Huber conducted very few experiments outside the hive, one of which though would have most likely allowed him to describe recruitment.

The question,

> […] to determine whether the scent of honey or appearance of a flower advertised its presence, I needed to hide the substance in a place that could not be seen. […]. I took small boxes of various sizes and colours, set small trapdoors of paper across holes in their lids, covered the floors with honey and placed them two hundred paces away from my hive. After half an hour bees arrived at the boxes, which they circled about and soon found the doors where they could enter; they pushed back the doors and make their way to the honey. [122, p. 196f.]

Had Huber monitored the boxes over a longer period, and because he had placed his boxes at a waggle dance distance of 200 paces, he would have certainly been able to record recruitment. Maybe, with a helper at the hive, also discovered the waggle dance.

Perhaps Francois Huber and Johann Ernst Spitzner sat at the same time, unknown to one another, in front of their hives. They would have been a dream team of early bee research.

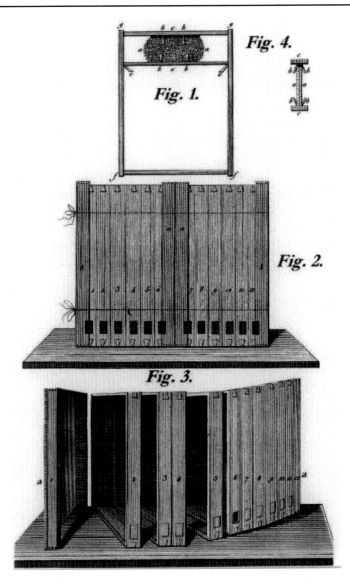

Fig. 2.4 The blind bee researcher Francois Huber designed a hive in which the combs could be turned like pages in a book

Christ Describes Clouds of Bees

Pastor Johann Christ (1739–1813) observed buzzing flights and recruitment. In his book from 1783, he writes:

> The evident excitement of a bee when she finds a dish of honey or makes some other discovery is testified by the emission of an unusual tone produced by her wing movements; this causes many other bees to gather rapidly and fly with her; soon a whole cloud of bees assembles and collectively exploit the find. [31, p. 36]

Christ observed that a bee could "cause" the assembly of other bees (recruitment) and a "whole cloud of bees" could work together (Buzzing flight).

Unhoch Looks Closely at Bee Dances

With his experimental approach, Spitzner was ahead of Nikolaus Unhoch (1762–1833) who nearly 50 years later also published a book about honeybees [270]. Nevertheless, Unhoch delivered the first useful description of the dance behaviour of honeybees although it remained a complete puzzle for him. In §13, entitled "Of the bee dance," he wrote:

> For some it would seem ridiculous, even unbelievable when I suggest that bees, when the hive is in good order, enjoy feelings of joy and well-being and indeed, in their own way, celebrate by performing a dance. I have often seen this and any who have a beehive with glass walls can watch the fun with their own eyes and convince themselves. A single bee barges into a cluster of bees standing peacefully by, presses her head down onto the comb, stretches her wings out and trembles her raised abdomen for a short period. The surrounding bees do the same, press their heads down and finally all turn together through slightly more than a semicircle first to the right and then to the left, five or six times in a round dance. Suddenly the dance leader leaves the group, joins another group of quietly gathered bees on the other side of the comb, repeats the performance, and the bees dance with her. The dance leader visits five or six groups, one after the other in different areas and on different sides of the comb. I have shown this to many bee friends who were amazed and laughed heartily. I observed the dances most often on fine warm days and in good hives. One will not see the dances on dull days, in hives that are in poor condition, or without a queen. I am not yet able to explain what these dances mean. Whether the dancing is an expression of general well-being or has some other as yet unknown purpose will become clear in the future, and with this bee ballet I close my first volume. [270, p. 115f.]

Bonnier Observes a Bee Accompanied by Another

The French bee researcher, Gaston Bonnier (1853–1922) described in detail, among other bee studies, his observations on the arrival of forager bees on flowering branches of buckthorn, a species of nightshade. The question that Bonnier addressed related to the interaction between searchers and foragers. He recorded the arrival time and the number of bees landing on buckthorn flowers. In a series of experiments, he cut twigs of flowers off and transported them to a new location. He recorded that on the following day a searching follower found the relocated flowers. Bonnier marked her with a spot of paint. The bee flew with nectar and some pollen back to the hive and returned to the buckthorn after five minutes, accompanied by another bee [20]. Considering these observations, Bonnier had no doubts about communication and recruitment.

Maeterlinck Experiments with Recruitment to Food Sources

Maurice Maeterlinck (1862–1949) made observations similar to Bonnier but took a different approach. He was one of the early bee researchers whose experiments and thoughts about the lives and communication of honeybees were full of original, basic ideas and interesting observations. It pays to examine his work carefully. Marking individual bees with a small spot of coloured paint—Maeterlinck and Bonnier were the first bee researchers to use this method—was decisive in probing the secrets of honeybees.

When tracing the origins of the present established view of recruitment in honeybees, one is confronted by an amazing reality: Maeterlink's discoveries and thoughts have, to the present day, not been adequately acknowledged, nor valued, or so incorrectly cited that exactly the opposite impression prevails.

Maeterlinck was not only a thorough observer of Nature but also a gifted writer, honoured in 1911 with the Nobel Prize for literature. These talents led to the excellent and factually accurate text, *Das Leben der Bienen* (Fig. 2.5) from which the following excerpt is taken:

> In the end, in order to obtain a more precise image of their intelligence, we need to determine how they communicate with one another. That they communicate is as clear as day; a community with such a large population that carry out such a diversity of activities in admirable harmony could not exist without the ability of members to emerge from their solitary intellectual state and interact with one another. They must also have the ability to express their thoughts and emotions, be it through speech, or more likely, a tactile or magnetic transmission that couple material and senses that are completely unknown to us. These senses could be located in their mysterious antennae with which they palpate their way through the darkness and which those of worker bees, Cheshire has calculated, consist of twelve thousand touch fibres and five thousand odour cavities. That they not only communicate about their routine activities but that their language also has expressions for the unusual can be assumed because news, good or bad, normal or unnatural, for example, loss and return of the queen, invasion of an enemy, a foreign queen, approach of a robber swarm, discovery of a treasure (the initiation of recruitment to a food source, the author) and so on, spreads through the hive. The behaviour and tones bees emit are so different and characteristic for each eventuality that it is not difficult for the experienced beekeeper to guess what is happening in the busy darkness of the hive.

> Should one wish to have clearer evidence, then one must observe a bee that has discovered a few drops of honey on the windowsill or tablecloth. First, she feeds so greedily that one can mark her with a small spot of coloured paint without disturbing her. The greediness is

only apparent. The honey does not reach her, so to say, personal stomach, but remains in a honey stomach, which is a sort of community stomach. As soon as this is filled, the bee flies off but is not blind and like a fly or a butterfly. Instead one sees her fly backwards, then attentively around the windowsill or table cloth with her head facing the room. She imprints in her memory the area and exactly where the treasure lies. She then flies back to the hive, empties her haul into a storage cell and in three or four minutes appears again at the wonderful windowsill to collect another load. She returns every five minutes, as long as honey is there, even if it lasts until evening. She flies back and forth without granting herself a moment's rest, from window to hive and hive to window.

I will not embroider the truth, as many have that written about bees.

Observations of this kind are only of interest if they are completely honest. Perhaps I would have said that bees are unable to communicate about events outside the hive if I did not, after an occasional experimental disappointment, find the pleasure to again establish that basically, humans are the only beings gifted with reason on this earth. [...]

I will also confess that marked bees frequently return alone. [...] Nevertheless, often enough, the fortunate bee returns with two or three companions. [...]

Once I marked the abdomen of a small Italian bee with a spot of paint. On her second visit, two sisters accompanied her. I captured both of these. She returned on her next visit with another three companions that I also captured, and continued with this until at the end of the afternoon I had caught and held eighteen bees. The original forager made eighteen of her sisters aware of the information.

Repeating the above experiment, one notes that the accompanying friends that have the solution to the happy find, do not always arrive together and that often there is an interval of many seconds between the arrivals of the individuals. One must raise the same question about information content, that Sir John Lubbock [Lubbock 1882 [159], J.T.] solved for ants: Do the companions of the bee that first found the treasure simply follow her, or are they perhaps sent by her and find the location by following her instructions and description of the area on their own? Were this the case, it is not difficult to see that there is a vast difference with regard to the extent and range of information content required [...]

My workroom in the country lies above the ground floor. Bees do not fly particularly high other than when the chestnuts and linden trees blossom, so that a piece of uncapped honeycomb (i.e. comb filled with honey from which the wax covering had been removed) lay for more than a week on the table before a single bee, attracted by the scent, flew to the comb. I took an Italian bee from a nearby observation hive, carried her up to my office, and let her feed on the honey while I marked her with a coloured spot.

When full, she flew back to her hive. I followed and saw how she hurried over the bees, buried her head in an empty cell, deposited the honey and made her way to fly out again. I repeated this experiment twenty times with different bees and each time captured the lured bees so that others could not follow their trail. To achieve this, I placed a small glass box, divided into two compartments with a trapdoor, at the entrance to the hive. If a marked bee came alone out of the entrance, I captured her and then waited in my room for her friends to whom she had brought the information. If the marked bee came out of the hive with two or three others, I held them captive in the first compartment and separated them from the marked bee. These I then marked with a different colour, released them and watched them. It is clear that if auditory or magnetic information had been provided describing a location, or orientation information to be employed on the way there, then bees set on the path should have been found in my room. I must confess that only one bee appeared. Did she follow the instructions given in the hive or was this pure chance? The observations were insufficient and circumstances prevented a continuation. I set the captured bees free and soon my workroom was filled with a humming crowd that had been shown the way to the treasure in the usual way. [160, pp. 92–98]

Maeterlinck was extremely cautious in the interpretation of his results, although the tendency was clear. Single experienced bees appeared at the feeder accompanied by two or three recruits (up to 18 recruits, added over a number of flights of the experienced bee). Alternatively, when the experienced bees were prevented from flying to the feeder, Maeterlinck, in all of his observations, detected only a single bee at the feeder. The importance of an experienced bee's flight to the feeder for recruitment is clear, nevertheless Maeterlinck questions the result because one bee found its way without an experienced bee.

Fig. 2.5 Title page of Maeterlinck's bee book

Do the companions of the first bee to discover the treasure only follow her or are they perhaps sent by her and find the way on their own according to the description of the site? [160, p. 97]

A combination of the two hypotheses did not occur to him but he certainly recognised that the matter was more complex than either one of the two alternatives, when for unknown reasons, he had to break off the study because "circumstances" did not permit their continuation.

Research on Bees Flourishes with Karl von Frisch

3

With his carefully planned experiments, Karl von Frisch initiated one of the most intensively studied areas in modern zoology. An end to the unanswered questions is not yet in sight—just the opposite.

Von Frisch Discovers That Forager Bees Exchange Information About Food Sources

Fundamental experimental research on recruitment in honeybees began with Karl von Frisch (1886–1982). In his early research, he focused on olfaction [70] and colour vision in honeybees [69]. During these studies, which at first had nothing at all to do with recruitment, the keenly observant von Frisch noticed a phenomenon that would lead him to the recruitment theme.

In 1923, he wrote (in relation to earlier studies):

> An unexplained observation that I repeatedly made in earlier studies on colour vision of bees, would not leave me in peace and demanded a closer examination. I saw that a honeycomb set out in the open lay for hours, even days without being discovered, was then found by ONE bee, and that others very soon followed. Their number grew to dozens in a few hours. I could only confirm and be content with the local opinion that successful foragers are noticed in the hive and that other alerted bees follow them on their return to the source. ([72, p. 9], emphasis in the original)

© The Author(s), under exclusive license to Springer Nature Switzerland AG 2022
J. Tautz, *Communication Between Honeybees*,
https://doi.org/10.1007/978-3-030-99484-6_3

The focus of this earlier investigation, published by von Frisch in 1914, was on a group of foragers he had trained to a particular feeder. To distinguish these from other untrained bees, he marked them with a spot of paint. In 1923, von Frisch wrote the following about this experiment (and cited again the text from his 1914 publication):

> There is no doubt that they (the marked forager group, J.T.) remain peacefully in the hive when thousands of foragers loaded with pollen and nectar returned home from visits to flowers. Why were they then mobilised when a single bee, THAT BELONGED TO THEIR GROUP, came home? Out of necessity I assume, 'that out of the many thousands of occupants of the hive, the few individuals that visit a particular feeder, are continually in touch and in in a certain sense know one another personally'. ([69, cited in 72, p. 10], emphasis in the original)

Von Frisch Realises Dances Encourage Inexperienced Bees to Fly to the Advertised Site

In the end, this observation was the basis of the dance language research program: Bees are not only encouraged by foragers to visit places they are familiar with; they appear at sites previously unknown to them from which the foragers return home.

The precedent was an observation brought to light by marked, trained bees in the hive, described by von Frisch as follows:

> Some of the bees – sometimes all – that trotted after the dancer were UNMARKED and did not know our feeding site. ([72, p. 34], emphasis in the original)

In the same text, von Frisch wrote about such newcomers that turned up at the artificially set-up feeders:

> It is generally accepted, and at first I was convinced, that a bee that had discovered a new food site LED her hive companions to it through repeated flights. ([72, p. 92], emphasis in the original)

Simultaneous observation of behaviour of bees in the hive and at the feeder directed the attention of researchers increasingly to dances within the hive, because:

> As long as the honey flow is strong, the returning bees dance in the hive; and as long as they danced, so new arrivals continued [new recruits, J.T.]. [72, p. 93]

This observation finally defined the problem area that has intensively engaged biologists for almost a hundred years: The dance language of honeybees.

Von Frisch Discovers That the Flower Scent on a Dancer Lures Bees to a Site Familiar to Them (But Not to the Goal of the Dancer)

During his study of the role of scent in the search for feeders, von Frisch and his co-workers performed an experiment in which a dancer was scented by a flower different to that from which she came and advertised in her dance. These "scent" followers visited sites known to them and associated with the particular scent. Bees remember the appropriate site; they do not follow the dance figure and consequently do not fly to the site of the dancer [84].

A citation from the study of von Frisch and Gustav Adolf Rösch:

> The round dance of a sugar water forager carrying pollen on her hind legs, or whose body is powdered with pollen from roses, alerts the rose pollen gatherers. (These fly to a site different to that from which the sugar water forager came; [84, p. 20])

Johnson [129] soon confirmed that the scent of a particular flower detected by experienced foragers in the hive could result in their visiting a site known to them characterised by the same scent. That foragers, noticing the scent carried by a dancer, did not fly to the site of the dancer but to a site the forager associated with the scent and did not receive the attention of bee researchers for more than eighty years after the studies of von Frisch and Rösch. Up to ninety percent of dance following can lead to a reactivation of experienced foragers already aware of food sources, they then visit [16, 109, 111, 112, 116].

Using highly modern techniques, the first to capture the behaviour of a forager aware of a feeder, but followed a dance to another site, was Rodrigo De Marco. He recorded the behaviour of foragers during their following of a series of dancers in an observation hive as well as tracing their subsequent flight paths with radar. Figure 3.1 provides an example of the flight path of a forager that had followed dozens of dance rounds of a dancer but then ignored the directions given by the dancer.

Additional results and ideas in relation to the circumstances under which the directional information in the dance is employed or not by the dance follower are not within the scope of this book (however, see [11, 44, 45, 62, 92, 110, 112, 148, 240]).

In general, a dance follower has a number of options in finding the goal subsequent to following a dance, but not from the information contained in the dance.

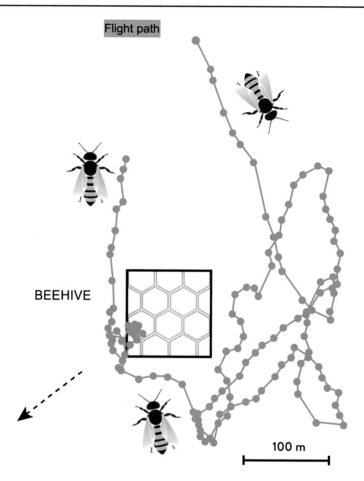

Fig. 3.1 The flight of a forager about six minutes after she had followed a dancer for dozens of rounds. The dancer advertised for a feeder that lay southwest of the hive (arrow). The bee flew north and returned to the hive after half an hour. The time intervals between points represent five seconds (First publication of such a flight path. Rodrigo De Marco, personal communication)

How von Frisch Perceived Recruitment to a Goal

The idea of the dance language in 1923 still lay in the future. At the start, von Frisch formulated his research on recruitment as follows:

> Should a bee find the feeder again filled after a long pause, are her colleagues fetched or sent to the feeder? DO THEY RECEIVE A SIGN FROM THE FORAGER THAT THEY SHOULD FOLLOW HER; OR A SIGN THAT THE FEEDER IS AGAIN FULL AND THEN THEY FIND THEIR WAY INDEPENDENTLY? One could get a clear answer to this question if fewer customers arrived at the feeder and at greater intervals, after a long pause in feeding. ([72, p. 27], emphasis in the original)

In relation to the foragers that already knew the feeder, von Frisch expressed a possibility that had already occurred to Maeterlinck. Both suggested two alternatives for successful recruitment: Either the recruits followed the experienced bees to the unknown goal or they flew there on their own. These two hypotheses were the guidelines for the following research about which Maeterlinck wrote:

> […] nothing more than that they follow her or are perhaps sent and find it themselves from her instructions and description of the site? [160, p. 97]

Another possibility—to start thinking in principle about a combination of these two alternatives—was not considered by von Frisch. With this premise, he limited his research to deciding between two alternatives. This had consequences.

Von Frisch Investigates if Bees Mark the Goal with Buzzing Flights

During an earlier study of the recruitment of honeybees, von Frisch made a highly significant discovery at the feeder. He observed a particular behaviour of experienced bees through which the newcomers reached the goal: The experienced bees scent-marked the site.

Initially, von Frisch assumed the mechanism of site marking to be the unusual "buzzing flights" that can be noticed at the feeder. In 1920, he was nevertheless sceptical of buzzing flights playing a role in communication between bees. He wrote, in the *Bayerische Bienenzeitung*, about the discovery of Spitzner who he admired as the "most astute bee observer of all time":

> I am most thankful to Dr Manger for drawing my attention to some old observations that I would have had difficulty finding. I cannot understand why such observations are completely forgotten and not mentioned anywhere in the newer literature. In relation to the "high frequency tone of joy", I will not deny that a bee flying rapidly from hive to feeder emits a different (high-pitched) tone to that of a leisurely wandering bee. That this difference is meaningful as a communication channel is a different question. [71, p. 168]

Von Frisch's interest in the recruitment phenomenon in honeybees awakened and he returned to the old mention of the "high-pitched tones of joy" and began a systematic investigation. He described the buzzing flights as follows:

> The flight tone of a bee that has found a rich food source is indeed somewhat special. This occurred to me first as I fed two groups of bees at the same time, one from a full dish of sugar water and the other from blotting paper. The flight tones of the approaching bees from the two groups were clearly different […]. [72, p. 150f.]

Von Frisch carried out an experiment to test whether the striking high-pitched flight tones emitted at rich food sources could attract other bees. He had the first apparatus for the study of bee behaviour built. A tuning fork with adjustable

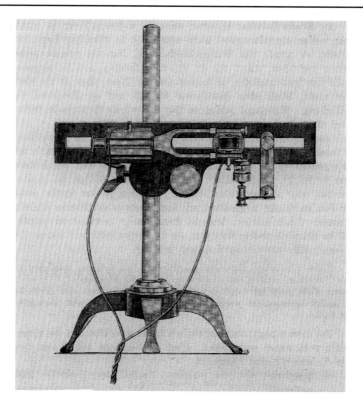

Fig. 3.2 The first apparatus developed for the scientific investigation of bee behaviour. Karl von Frisch used this electrically activated and adjustable frequency tuning fork to test the hypothesis that the tone of the remarkable buzzing flights could lead recruits to the feeder (from [72])

resonant frequencies, driven by an electromagnet (see Fig. 3.2), was designed to imitate the tone of buzzing flights. It appeared to work splendidly, and bees soon surrounded the buzzing tuning fork. However, when von Frisch switched the tuning fork off, he obtained the same result with bees surrounding the object. It was not the tone that had attracted the bees. Instead, "the shiny instrument awakened the interest of bees flying around in the same way" [72, p. 156].

All bee researchers face this problem in field studies: What stimuli do experimental setups and attending researchers represent for bees at sites where their arrival is studied. Is the behaviour of bees influenced? Researchers could unwittingly provide cues to searching bees that more or less falsify the results of the study. Only carefully employed control experiments can solve this problem.

Von Frisch Investigates Scent-Marking

After he had shown experimentally that it was not the flight tone that attracted recruits to the feeder, von Frisch turned his attention to another difference between bees approaching a rich or a lean food source: the visibly everted scent glands on the abdomens of bees (Nasanov glands) (see Figs. 11.3 and 11.4). The single citation from von Frisch's publication of 1923:

> [...] it is not unusual, during an extended period of rich feeding to see that ALL bees in a group evert the scent organ at EVERY approach. At poor sources, however, even during prolonged foraging, the organ is NEVER everted. ([72, p. 160f.] emphasis in the original)

Further, after simultaneous observations at the hive and feeder:

> Only after they had returned to the hive and danced on the comb and then returned to the feeder, did they evert their scent organ. [...] Only now, I noticed that they everted the scent organ most often during their flight around the feeder, impregnating the area around a rich food source with a particular scent. Finally landing at the feeder, they evert the scent gland again for a short time. [...] The longer a bee danced on the comb, the more intensely she courted the attention of recruits, the more energetically she scented the site on her return. [72, p. 161]

Relating to communication in the field, von Frisch noted:

> Thereby it is shown that the normal strong stream of new recruits to the feeder is in the main determined by the eversion of the scent organs by the foragers. [72, p. 166]

Further:

> Foragers exploiting a rich food source swarm for a long time around the site when they arrive from the hive, with scent organs everted and also when they are feeding, thus impregnating the area and the food source with a specific odour. By this means they lure circling and searching recruits from a wide area and lead them to the correct site. [72, p. 175]

The role of scent organs of honeybees in leading recruits to a goal was again thoroughly studied and discussed in a detailed publication in 1926 [84].

Von Frisch discovered the following in his last original study concerned with communication between bees in the field:

> When a forager had danced and returned to the flowers, she hovered for a half to one minute around them with an everted scent organ. She kept the organ everted at the beginning of nectar collection from the first or second bloom, exactly as when she approached a sugar water feeder: at first during circling flight with the organ everted and only withdrawing it when feeding. When flowers had been repeatedly visited and accumulated nectar taken, the now lean source was still exploited but without dances in the hive and scent organ eversion when approaching flowers. [84, p. 4f.]

Further:

> Contrary to certain doubts, it was determined that bees also employ their scent organs for communication when harvesting nectar and pollen from flowers, as when collecting sugar water from a feeder. [84, p. 20]

In the middle of the 1920s, it was clear to Karl von Frisch: Experienced bees danced in the hive, aroused the interest of dance following bees that there was something worth collecting outside the hive, and then marked the goal for the searching recruits that she attracted over a wide area. The goal was marked with the pheromone contained in the Nasanov glands located in the abdomen (see Fig. 11.4), the main component of which is geraniol, which smells like geraniums.

Julien Francon Describes Goal Marking Without Defining It

Experiments conducted and published in his book in 1938 by Julien Francon supported the marking of a goal by experienced bees [65]. Nevertheless, his observations were not described in this way. For him, his experiment provided puzzling proof for the "intelligence of the bees."

The precedent was an observation that was not new to bee research. Francon trained bees to a feeder (see the Appendix for a method to train bees) and marked them with coloured paint so that they could be recognised. He, like so many others before him, experienced the following surprise:

> A few seconds after their (the trained and marked bees, J.T.) fifth return, the great occasion: an unknown bee, after eventually landing, joined them! [65, p. 49]

These studies led Francon to undertake a series of different experiments. A particularly striking one consisted of a hidden feeder to which he built two entrances. One entrance was easy to find, the other difficult. Recruits that arrived without the company of trained bees always chose the entrance used by the trained bees. They took the difficult path, a long tunnel hidden under leaves if the trained bees had previously used this.

Here, Francon's description and interpretation, in which he speculates that instructions to recruits he observed, are so precise that the correct of the two alternative entrances, partly hidden beneath stones, was indicated (By "baiting bees" Francon means bees that were trained to a well-hidden feeder, and as "helpers" the recruits):

> 1. Untrained bees (helpers) pay little attention to feeders, the worth of which they are unaware.
>
> 2. Trained bees (baiting bees) can only find the feeder again if they have learned the way there.
>
> 3. Once the trained bees know the way to the feeder, they will find it again even under difficult circumstances.

> We have established many times that recruits (helpers) find their way on their own, without any help, to feeders, the worth of which only the trained bees are aware. In addition, these recruits, that otherwise were not able to find an entrance they had not noticed before, were as good as the trained bees that discovered the unmarked way to the feeder. The invited bees would have without doubt not enjoyed the small piece of sugar had they not been informed precisely of its worth, its location and the complicated way to it. Only the trained

bees possessed this information at the start of each study. It follows that it was they who informed the recruits and shared the necessarily exact and reliable instructions with them as all of our described tests showed. [65, p. 176f.]

Francon did not comment on how he imagined the recruits were informed. Finding the hidden entrances to the feeders suggests chemical marking of these by the trained bees. This did not occur to Francon. His confirmation of a "completely unmarked way to the feeder" would not withstand testing with modern chemical methods.

Buzzing Flights and Scent-Marking of Goals Are Forgotten

The discovery of buzzing flights around feeders, scent-marking, and its attraction of recruits was not systematically investigated in later studies of von Frisch and his students, nor mentioned in the interpretation of experimental results with the exception of the following:

After the development of the idea of the dance language, von Frisch returned once more to the experiment in which he had shown the attractive role played by geraniol. This time with the aim to test whether the discovery of the feeder by recruits took place in the absence of geraniol. To this end, he sealed the Nasanov glands of experienced bees with shellac. Because he recorded a similar recruitment success as with untreated bees, he concluded that geraniol was not an attractant and that the dance was responsible for the discovery of the feeder.

Critics were of the opinion that these experiments were unconvincing for two reasons: without the appropriate chemical methods to test how completely the glands were sealed, it could be that some geraniol had in fact escaped. Von Frisch believed that the closure of the glands was total (in relation to the technical problem of actually achieving "complete closure" of parts of bee's bodies, see p. 50), but could not determine this without the chemical analysis. The second criticism concerned the strong odour of shellac that could have itself acted as an attractant and influenced the experiment.

Be as it may—bees treated with shellac still perform buzzing flights at the food source.

Buzzing flights and open Nasanov glands continued to be included in the ensuing experiments of von Frisch. Nevertheless, not as part of the studies. Quite the opposite, namely as factors he purposely excluded and emphasised in his publications. He mentioned repeatedly that he purposely did NOT pay attention to events at the feeder because here experienced bees attracted recruits to the goal.

In his famous step and fan studies, he consistently studied events at artificial control stations not visited by experienced bees and excluded feeding sites frequented by experienced bees. (compare Figs. 3.3 and 3.4).

Von Frisch Perceives the Dance to Contain Information About Locality

Thursday, 10, Friday, 11, and Saturday, 12, August 1944 were, in his own opinion, remarkable dates in the research life of Karl von Frisch described in an extensive publication in 1946 [75]. On these three days, for the first time, he set up feeders to which different groups of marked bees had been trained, and control stations at which recruits could be counted, at different distances (10 and 150 m; 15, 140, and 150 m, respectively) from the hive.

Von Frisch himself was at first not clear why he had set up this experiment or what he expected. It could be that some casual observations from his circle of students motivated him. They had seen that recruits did not necessarily visit the food source nearest the hive if other feeders at various distances had been set up. Here again is another example: An imaginative, unbiased, and playful attitude to research themes often leads to striking new discoveries.

Although all stations were scented with lavender, most recruits arrived at those feeders experienced bees visited. This confirmed the experiments that von Frisch had published in 1923.

New and totally surprising, because completely unexpected, only a few bees landed at control stations close to the hive. Instead, most recruits flew to more distant control stations [75]. Only previous activity of forager bees could be responsible for this.

The level of understanding of von Frisch and others from his previous experiments at that point in time is cited from his publication in 1946:

> At rich food sources, bees visiting flowers and also sugar water feeders, evert their scent organs, glandular pockets in the integument near the tip of the abdomen, at the goal and can lead recruits that are searching NEARBY to the correct goal with the scent organ's intense attractant perfume. (p. 2, emphasis J.T.)

In 1923, von Frisch formulated the same situation:

> [...] DO THEY ATTRACT THE SEARCHING RECRUITS THAT ARE CIRCLING AROUND IN A WIDE AREA? ([72, p. 175], emphasis in the original)

Similarly, one year later:

> The well fed bees evert their (SCENT ORGANS), glandular pockets located near the tip of the abdomen that extrude an odour perceptible to humans, when they approach the feeder and also during sitting and drinking at the feeder. In certain experiments I could show that this odour is intense for bees and has an effect over a CONSIDERABLE DISTANCE [emphasis J.T.]. Bees flying to poor food sources never evert their scent organs. It is the odour from these scent organs that attracts the recruits from significant distances to the site [...]. [73, p. 23f.]

No experiments, measurements, or data are known that von Frisch used to initially assume attraction over a large distance, or to correct this later to "from a wide area" and over time to "nearby." However, these changes in the estimation of the range of the attractant, based on assumptions, decisively changed the direction of subsequent research.

This is emphasised in the following citation from a publication of von Frisch that appeared in 1965:

> Neither the odour from scent organs of foragers released before landing, nor the scent of visited flowers, can build a bridge between a distant pasture and recruits in the hive. Only when they are steered to the area of the food source with information about distance and direction, can the odour from the scent glands be effective. The scent gland odour acts in the immediate vicinity of the successful foragers [...]. [78, p. 229]

What von Frisch describes here in his words corresponds to the concept of a three-phase mechanism of distant goal orientation that ornithologists first formulated as the result of research on bird migration [277, Fig. 1]. Presented here are SEND (directed) and ATTRACT (lure), as the first and the last phases of the behavioural chain. The intermediate phase, the search phase, and transition between SEND and ATTRACT, in which goal orienting stimuli and signals are present, are hidden "nearby." However, what is meant by "nearby?".

"Nearby" is not a clearly defined distance and is of little help. A statement is needed about where, during the flight from the hive, the approximate information in the dance is transformed into direct goal orientation; where along the flight from the hive to the food source are Nasanov glands of foragers opened and on what external factors is this dependent.

We still know nothing about the search phase and transition to attraction. In the research, the best we can find is a generalised formulation such as that in the above-cited text from von Frisch.

In the absence of adequate techniques, so far the only way to decide if their scent glands are open or closed has been to observe slowly flying bees. This simple way of estimating the employment of the Nasanov glands by bees is used to this day in bee research.

The view that forager's scent organs are effective only in the proximity (and here close proximity) of the feeder is also shared by Adrian Wenner [284]. Here too, in his study, no details of experiments are provided to support the determination that attractant from the scent organ is limited to "near" the food source (what is "near?").

Assumptions about the employment of scent organs by worker bees have not been followed up in any systematic investigation. This is a neglected area in the communication research of bees. Here again, we come across the blind spot, the missing half in the study of communication between the forager bees in the field.

Von Frisch Understands the Dance Figure Relates to the Food Site

In bee research, time and again, the choice of words resulted in a change of direction, including when this relied on assumptions instead of observations and data. When "from a wide area" is expressed later as "nearby," without data for one or the other, the expectations and picture that one has of the events change. So too, the meaning of an observation on the discovery of the food site by recruits. If one believes that the scent attracts only those in close proximity, one must then look for that which brought the recruits this far.

This is what von Frisch did and found the answer in the bee dance. His publication in 1946 contains the pivotal discovery of the existing bee research, which von Frisch expressed thus:

> The distance to the food source from which the dancer brought her booty can be accurately determined with a timer and the number of turns she makes. In addition, her waggle dance also indicates the direction of the food source. [75, p. 48]

This discovery, overlooked by all his predecessors, was justifiably recognised as one of the most important in modern behavioural biology and one may assume was the most important single finding that led to the award of the Nobel Prize to von Frisch.

With this, although the distance was cautiously formulated ("*accurately*"), the attention and emphasis of research and interpretation of results of a large community of researchers in the following decades focused firmly on events in the hive.

Karl von Frisch brought his interpretation of recruitment again to the point in his well-known book *Tanzsprache und Orientierung der Bienen* [78]:

> Foragers that visit a good food source quickly bring many recruits even to distant and hidden sites. These are NOT LED TO THE GOAL but SENT. ([78, p. 236], emphasis J.T.)

Twenty years earlier, von Frisch was not so certain about the overriding importance of the "sending." In his publication of 1946, he writes:

> Is it not the scent from the foragers alone that causes a large number of recruits to appear at other observation sites near the feeder? [...] The odour from the scent organs of the busy trafficking bees may spread over the entire area and hence include the neighbouring observation sites. [75, p. 9]

To clarify the situation, he developed the idea for his later important fan experiments.

> In order to examine this with further feeder experiments at distant locations, I set up observation posts at the SAME DISTANCE from the hive but IN DIFFERENT DIREC-TIONS, outside the scent range of the foraging bees. ([75, p. 9], emphasis in the original)

Here too can be noted that von Frisch had reached a conclusion about the range over which the scent of the forager bees was effective for which he provided no data.

The critical experiment that led von Frisch to his clear statement ("These are not led to the goal but sent to it") emerged from the systematic development of his ideas in his studies in 1944 in the form of the step and fan experiments.

During the fifties, von Frisch tested his idea that bees transmit details about the food site by way of a dance language, with his renowned step and fan experiments. The aim was to establish if the recruits in fact followed information about a goal contained in the dance. The step experiments should reveal if the information was provided about distance and the fan experiments should demonstrate that the direction of the site was understood.

With Step Experiments von Frisch Investigates if Bees Convey Information About Distance in Their Dance

Let us first look at his step experiment of 10 July 1956 (from [78]). A small dish containing sugar water, scented with orange flower oil (orange in Fig. 3.3), was set up 1050 m from the hive along a path. (See Fig. 12.10 for the paths along which such experiments were conducted.) Thirty forager bees were marked with colour and trained to this site. (See Appendix for the method used to train bees to a feeder.) At first, bees were provided with an unscented and low concentration sugar solution. The bees collected the sugar water but did not dance in the hive for this diluted sugar. To begin the experiment, the concentration of sugar was increased and bees began to dance in the hive. The sugar solution was now also scented. Trained bees that now flew back and forth from hive to feeder recruited 132 newcomers during the three and half hours of the experiment, all of which landed and were captured. Eight control stations were set up along the line between hive and feeder. These were set up exactly as the feeder and scented but contained no sugar water. This should determine how closely recruits keep to danced information in relation to distance to the feeder. The following data were obtained:

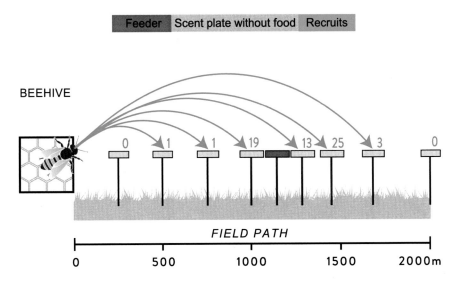

Fig. 3.3 One of many step experiments that von Frisch and his collaborators carried out. Thirty foragers were trained from an observation hive to forage at a feeder (coloured orange), 1050 m away. These bees danced in the hive and activated recruits. Eight control stations were set up in a line at different distances from the hive. Recruits approaching within 20 cm were counted (but not those that reached the feeder). The experiment showed that most recruits turned up at control stations closest to the feeder (number and flight path of recruits in blue). The figure is derived from the original in von Frisch's book published in 1965 [78, p. 90, Fig. 85]. For a general view of the conduction of these experiments, see Fig. 12.10

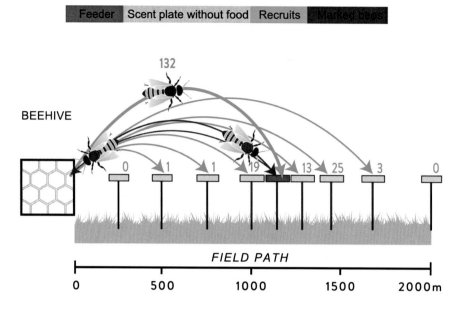

Fig. 3.4 For some of Karl von Frisch's step studies, it is possible to find out from the text describing the experiment how many recruits landed directly at the feeder (coloured orange) for which foragers danced and continued to visit during the experiment (red flight path). 132 recruits in this experiment reached the feeder advertised by experienced bees (numbers and flight paths of recruits in blue). If one transfers these values to Fig. 3.3, it appears that the recruits to the control stations lost their way

> An observer seated at each scented control station noted the number of unmarked circling and landing bees. [...] they arrive close to the ground against the wind, in a slow zigzagging searching flight to swarm around the plate. They were only counted if they came within a distance of about 20 cm. [78, p. 86]

Recruits that flew to control points during the three-hour observation period were registered as "scent plate arrivals" in the publication. All were counted which incurred the risk that those that did not land but only circled the site were counted more than once. This limitation appears in the literature as an argument against the credibility of these experiments (e.g. [283]), although this is not the real problem. If one adds the numbers of recruits, shown in Fig. 3.3, with those (the numbers of which can be extracted from the text in von Frisch's book) that flew directly to the feeder, landed, and were captured, so the "scent plate bees" appear as exceptions (regardless of whether they landed or not, or were counted more than once). With a total of 132, most of the recruits in this experiment landed at the feeder to which the experienced bees (the dancers) continued to fly and forage during the observation period.

A comparison of Figs. 3.3 and 3.4 leads to the following consideration: One could interpret the data from the step experiments in accord with the ideas expressed in this book and propose that recruits counted at the control stations had not made contact with experienced bees in the field or had lost it.

From step experiments, the following is clear: (1) By far the most recruits turn up at the site where experienced foragers are collecting. (2) More recruits arrive at control stations located close to the feeder than those situated further away.

Such an experimental result leads to two related questions that, depending on one's point of view, can be weighted differently and influence which aspect of recruitment to a feeder one wishes to give preference to at this stage.

The first question focuses on events at the feeder: What brings so many recruits to land exactly at the feeder visited by thirty experienced bees? This question leads directly to an investigation of the communication mechanisms employed by recruits in the hive and in the field.

The second question focuses on events at control stations: How do recruits distribute themselves over control stations that are scented but offer no food and are not visited by foragers? What led recruits to these stations? These questions exclude an investigation of communication between the experienced and inexperienced bees in the field and concentrate on events in the hive.

Karl von Frisch chose the second question and consciously ignored events at feeders visited by experienced bees and advertised in their dances. His comment on the enormous numbers of recruits that in all his experiments landed directly at the feeder and which he did not mention in later experiments is as follows:

> WITH THIS IT IS DEMONSTRATED THAT THE NORMAL STRONG STREAM OF NEWCOMERS TO THE FEEDER IS MAINLY DUE TO THE EVERSION OF THE SCENT ORGANS OF THE FORAGERS. ([72, p. 166], emphasis in the original)

Methods to study communication in the field between hive and feeder were not available at that time. It is, therefore, understandable that aspects of communication between bees in the field were neglected at an early stage due to the lack of adequate techniques. Nevertheless, this had a restrictive effect that prevails to this day, in relation to views on recruitment. Social behaviour out in the field was relegated to future tasks, ignored, and over time faded away as a research option—the blind spot is firmly established.

With Fan Experiments von Frisch Investigates if Bees Convey Information About Direction in Their Dance

The basic idea of fan experiments was the same as step experiments. A group of marked foragers was trained to a scented feeder. The recruits that arrived at control stations recorded.

Karl von Frisch presented results of fan experiments without information about events at the feeder (Fig. 3.5).

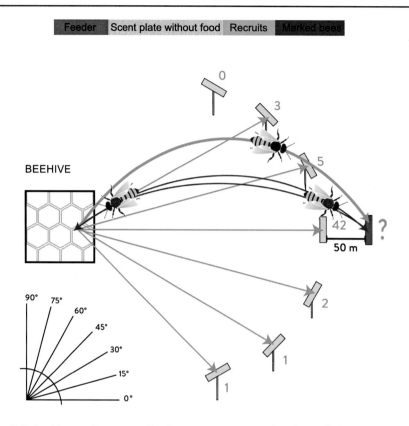

Fig. 3.5 In this experiment, von Frisch set up seven control stations, all the same distance (505 m) from the hive (the location of the stations is shown in the lower left of the figure). The appearance of the control stations was the same as the single feeder (coloured orange) but contained no food. The single feeder to which the experienced bees flew (red flight path) was located fifty metres beyond the control station in the centre of the fan. Bees arriving at control stations were counted. Most were recorded at the control station nearest the feeder (number and flight path of the recruits in blue). Recruits landing at the only feeder visited by experienced bees and captured there were not counted (after [78, Fig. 138, p. 159])

    As in step experiments, the nearer the control stations were to the feeder visited by experienced bees, the more recruits were counted there. In contrast to some step experiments, von Frisch provides no information about events at the feeder to which trained bees are recruited with dances and Nasanov gland scenting. There is also no information about the number of recruits that flew directly to the feeder and were captured.

    The role of communication between bees in the field, in spite of its fully appreciated and overwhelming importance, was consciously omitted in favour of examining the distribution of recruits over the control stations that carried the scent of the feeder but were not visited by the experienced bees.

A brief summary of the results of von Frisch's step and fan experiments remains: Dancers communicate the direction and distance of a goal. More precisely, to a site from which the dancer has come and advertises. The first figure in the literature that presents this statement about directional information in the bee dance is found, more or less modified, in thousands of publications (Fig. 3.6).

Directional information is never absolute, always in relation to a reference. For bees outside the hive, the sun provides the reference point, within the dark hive, it is the downward direction of gravity. Honeybees translate the perceived angle between their flight path and the sun, from the angle of the waggle phase of the dance, on the vertical surfaces of the combs, in relation to gravity. This sounds more precise than it is, as will be illuminated in this book.

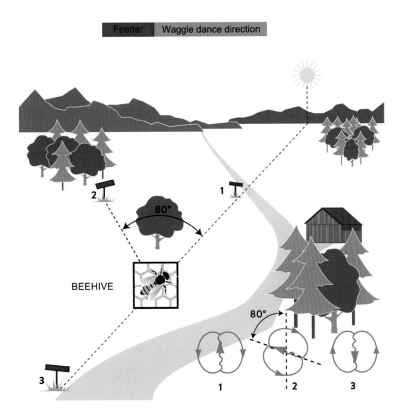

Fig. 3.6 The classical representation of Karl von Frisch's fan experiments revealing how dance followers decode directional information from the dance. Three dance figures are schematically represented, below at the right. Dances 1 and 3 indicate up or down and correspond to gravity. Because in the field, bees orient to the sun instead of gravity, the dance follower interprets the feeders (coloured orange) to lie in a direct line from the hive to the sun. Similarly, they find a feeder from dance 2 by flying 80 degrees left of the sun's position. The representation is after the original figure in Karl von Frisch's 1965 book [78, Fig. 119]

Von Frisch Was Awarded a Nobel Prize for His Research—The Alternatives

The step and fan experiments and derived conclusions about the communication abilities of honeybees were significant elements for the award of the highest distinction to Karl von Frisch that any researcher can achieve. Three behavioural biologists, Karl von Frisch, Niko Tinbergen, and Konrad Lorenz, shared the Nobel Prize for medicine in 1973.

The conclusions from the step and fan experiments, graphically represented in Fig. 3.6, are not without alternatives. At this stage, the research offered several possibilities as follows:

1. Recruits followed signals in the field given by experienced bees because a large majority of recruits landed at the feeder visited by experienced bees (the dancers). Recruits that did not land there never made contact with experienced bees. Von Frisch recognised this explanation in his first discovery of the recruitment mechanism (see p. 22).
2. Recruits follow information about the area from dances in the hive. This was the central message from the step and fan experiments. Formulated more cautiously one could also say: The waggle dances of foragers exert an influence on the distance and direction in which recruits fly in the field of control stations. The closer the control station lay to the feeder, the more recruits arrived there.
3. Recruits follow exclusively olfactory stimuli that arise from all stations. The dance contains no useful locational information for recruits and merely activates them to fly out of the hive and search [284].

Given the various explanations and possibilities, it is of no surprise that not all bee researchers were convinced of the functional significance of the bee dance as a dance language. In the sixties, a controversial view was expressed in which it was proposed that counter to von Frisch's opinion, dance followers derived absolutely no information about locality from the dance figures. The spokesperson of the largest group of bee researchers that supported this view was the US American biologist, Adrian Wenner.

The Dance Language Remains a Controversial Idea

4

> The debate about the information content of the bee dance is to this day not closed. For the community of bee researchers it is a true irritant – and has continued over decades. Tina Heidborn 2010 [117, p. 79]

Details of the experimental setup of step and fan studies, conceived and carried out by von Frisch, were varied by others and gradually led to results that could be explained without the communicative role of the dance. This led to the development of a perception of recruitment in honeybees that "threw the baby out with the bathwater."

Wenner Assumes That Bees Orientate Exclusively to Odour

Adrian Wenner addressed the question of whether recruits use the information given in the dances or ignore them and are guided to the goal exclusively by odour. He arrived at a radical conclusion including both correct and incorrect aspects.

In his opinion, bee researchers can indeed derive the location of the goal from the dance figure, advertised in the dance (see Fig. 3.6 for the direction of a goal) but not the dance followers. In his view, the dance contains no orientation information for the followers.

This is wrong. With the step and fan experiments, Karl von Frisch could show without doubt that flights of dance followers are influenced by the direction and duration of the observed dances. Wenner's hypothesis that the dance followers completely ignore the information about locality is incorrect.

Wenner's assessment of the dance language, though, is correct in that while researchers can calculate exact locations from dances of bees (see Fig. 3.6) and with a precision depending on the extravagance and complexity of the analysis and size of the database, this information and the complex analysis are not available to the dance followers. The dance followers are provided with half-truths and no precise information (cf. Figs. 3.6 and 7.3).

© The Author(s), under exclusive license to Springer Nature Switzerland AG 2022   39
J. Tautz, *Communication Between Honeybees*,
https://doi.org/10.1007/978-3-030-99484-6_4

Adrian Wenner and his co-author Patrick Wells thoroughly and in detail critically examined the von Frisch hypothesis of a dance language. Included in the
book, appropriately titled *Anatomy of a controversy* [284], are extensive philo-
sophical considerations of controversies and the scientific process.

In relation to the dance language, he and other researchers explained the arrival
of the recruits at the feeder solely through orientation to odour. He provides
extensive grounds for the perception that a connection between the details of the
bee dance and the location of the food source is meaningless for honeybees and that
the dance followers receive no indication about the location of a goal, [283]. In his
view, the dances merely encourage the recruits to leave the hive, initially in any
direction. According to Wenner, they reach the site advertised by the dancer
exclusively through olfactory stimuli in the environment. Employing an experimental setup similar to von Frisch with a single feeder and a number of control
stations (step and fan experiments), they interpreted the behaviour of the recruits to
be explained by their perception of a dominant olfactory point in an odour complex
arising from the interaction of all sources in the field [130, 284].

Wenner Versus von Frisch

A brief look at the concept of distant goal orientation (Fig. 1.1) shows that Adrian
Wenner's model for recruitment does not appear to fit. Phase 1, SEND. is absent,
assumed is that the behaviour of the recruits begins with Phase 2, SEARCH.
However, if, after the bees have followed a dance and been encouraged to leave the
hive and are then confronted by a head-wind with a strong olfactory signal, they
may well fly directly to the source (Figs. 4.1 and 12.1).

Karl von Frisch was aware of how important the scent of flowers is for bee
orientation. A considerable portion of his research on bee communication was
concerned with so-called scent guiding and the problem of how in agriculture one
could steer pollinating bees [76, 77].

Von Frisch wrote:

Farmers often wish that POLLINATION COULD BE INTENSIFIED. We have the possibility to achieve this. We know HOW BEES COMMUNICATE WITH ONE ANOTHER
and WE COULD LET THEM KNOW IN THEIR OWN LANGUAGE, WHERE THEY
SHOULD FLY. […]. If one brings a bee colony to a field of red clover and feeds the bees
sugar water with a taste of red clover flowers, these will dance in the hive, alarm their
colleagues and induce them, with the odour of clover, to fly to the field of red clover. ([76,
p. 3], emphasis in the original)

The "blind spot" established itself as the controversy between von Frisch and
Wenner intensified and focused entirely on whether the dance contained information used by bees for orientation or not [97]. A solution to the question of who was
right was seen as a keystone in research on bee communication [178].

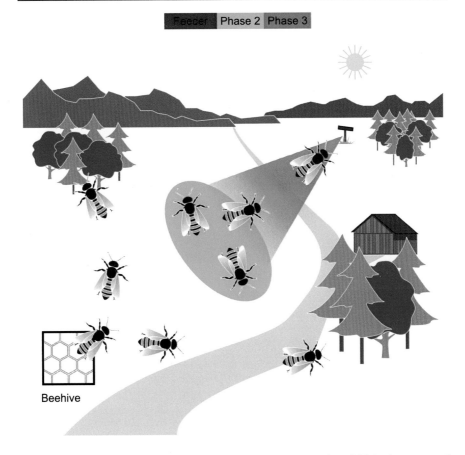

Feeder Phase 2 Phase 3

Beehive

Fig. 4.1 The results of the research of Adrian Wenner and his students laid the importance of odour from flowers for the discovery of the goal in the last phase of the distant goal orientation Phase 3, red sector (cf. Fig. 1.1). If wind brings the scent to the hive, the search field (green) moves close to the hive, or is no longer relevant (see Fig. 12.1). Distant goal orientation transforms to near goal orientation and the bees can fly directly from the hive to the goal

Lessons from the Controversy—An Interim Balance

What has already been written about the controversy will not be repeated here (see, e.g. [271] in relation to cognitive theoretical considerations). Wenner and Wells extensively expressed their criticism of von Frisch's experiments in their book [284]. The objections summarised there from the many publications of the Wenner school, extending over a considerable time, were in turn contradicted by the von Frisch school (e.g. [79]). The control experiments and controls of the control experiments, carried out by the two groups, and how each of these was interpreted

Fig. 4.2 Original experimental boxes (8 cm square) that von Frisch employed in large numbers to train bees to different odours. Scent sources that bees were required to visit were placed within the boxes [78]

by the two schools are not the subject of this book, because the controversy did nothing to enlighten the blind spot. On the contrary, focusing on one single issue among the many open questions led to the opinion that the settlement of the controversy would answer the question about how honeybees recruit.

Despite the award of the Nobel Prize to von Frisch tipping the scales against the Wenner hypothesis [117, 178, 179], the need to directly confirm the importance of the dance movements, independent of other goal orienting signals and cues, remained. The three possible explanations derived from the results of the step and fan experiments were equally weighted.

Wenner's view that the dance figures convey no information about the site was, then and now, shared by very few bee researchers. It proved to be wrong. The studies of Wenner and his co-workers nevertheless showed without a doubt (confirming the earlier studies of von Frisch, e.g. on scent guiding of foragers) how important the scent of flowers is for the orientation of the forager bees in Phase 3 of the distant goal orientation, when recruits, motivated by the dance, are underway searching for a source of nourishment.

Richard Dawkins made the mollifying comment on the controversy that one could reasonably assume several alternative possibilities being available to foragers in searching for food sources. Among others, these were the dance, scent, and the presence of other bees in the field. It would then depend on the circumstances which of these options dominated and were used by the bees.

Sixteen years later, Gould, Dyer, and Tyne expressed a similar view:

> In short, they argue (Wenner and co-workers. J.T.) the dance correlation is meaningless for bees that use only scent or a language for communication, because they provided a situation in which only scent could be used for recruiting. The error in this argument is, of course, the assumption that bees have only one way to communicate. [104, p. 142]

Wenner's view that bees could only communicate in the field (recruitment solely with scent) was justifiably rejected. The opposite was automatically assumed. It follows that if it is wrong that bees have only one of two possible mechanisms available to find a goal, then it should be correct to assume that bees equally use both discussed mechanisms. However, this is not the case.

The suggestion that foragers have several options from which they can choose to bring recruits to the goal is the most often presented in current research. Accordingly, all mechanisms are considered of equal importance and the option bees adopt depends on the prevailing circumstances. This would consolidate all observations and support one or the other views (see controversy).

This solution appears to explain the essentials but avoids the blind spot. It will be shown in this book that all is NOT explained, that all is NOT possible and all behavioural components are NOT equal, and that ALL mentioned behavioural components of bees are employed, nevertheless tightly integrated and not freely available as options for bees to choose from.

James L. Gould's Experiments Clarify Some, But Not All Difficulties

5

A study by the US American biologist James L. Gould aimed finally to resolve the controversy but not to research bee recruitment in all its aspects. The blind spot was ignored.

In 1975, during the most intense period of the controversy between the von Frisch and Wenner schools, Gould completed a doctoral study that had a significant impact on Wenner's criticism of von Frisch's work.

Tania Munz, a science historian, in the framework of whose doctoral study the *bee battles* between Karl von Frisch and Adrian Wenner was included, judged the importance of Gould's work [95] as follows:

> Many in the animal behavior community welcomed Gould's work as the final word in the controversy and hailed it as the vindication of von Frisch's theory. [178, p. 560]

With such high praise, Gould's work deserves a close examination. The data from Gould's work are taken here as they were published and considered within the theme of this book, namely the blind spot in relation to bee communication.

Gould himself was open and cautious in the interpretation of his results (see p. 54). The results of his work though were seen as the keystone in the persisting disagreements that had become even personal.

Taking the controversy about the bee dance as an example, Pokaloff expressed the danger of closed minds inhibiting new views as follows:

> The honeybee recruitment system has been debated extensively for many years. Perhaps what is needed are fewer debates and personal attacks and more observations and data. With well thought-out experimental designs, the communication abilities will unfold before us. Controversy is a human phenomenon, and it is time to take our personal feelings out of the equation and just watch the bees. [198, p. 193]

J. Tautz, *Communication Between Honeybees*,
https://doi.org/10.1007/978-3-030-99484-6_5

How Bees Orientate

The light outside the hive and gravity (in the hive) offer the crucial orientation parameters, as Karl von Frisch and his co-workers determined (summarised in [78]).

In the open, honeybees use the sun (and the polarisation pattern of the sky; [214, 215], see Fig. 12.12) as an essential orientation reference on their foraging flights. Gravity, in the dark hive, a vertically directed force, provides the directional reference in relation to which dances on vertically suspended combs are performed [163].

Honeybees have appropriate sense organs for the perception of these orientation parameters (Figs. 5.1 and 5.2).

Bees, as in many insects, employ a simple principle to detect the direction of gravity.

Small cushions of sensory hairs are located at the joints that connect adjoining movable sections of the body. If a freely suspended part of the body, for example the abdomen, the head, or the two antennae, is pulled downward by gravity, so the

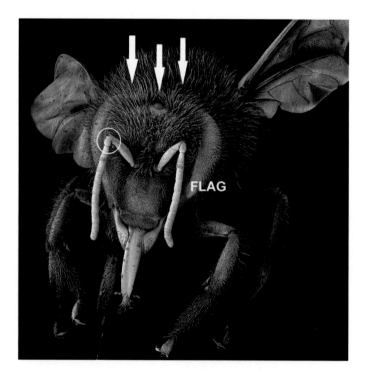

Fig. 5.1 The head of a worker bee. Their most important senses are vision, with two large compound eyes and three tiny single facet eyes hidden among hairs of the head (= ocelli, buried among hairs. Arrows show one, other two hidden), and an olfactory sense contained in the two feelers (antennae). Johnston's organ (circle), situated between short and long segments (flagellum), is activated when the antenna is bent by air movement

Fig. 5.2 The compound eye of a bee viewed with a scanning electron microscope. Each of the hexagonal facets forms an image. The facets directed upwards in flight can detect polarised light. This is particularly important for directing the flight of recruited foragers in phase 1 of distant goal orientation. Touch and air currents stimulate the small sensory hairs between the facets

receptors on the one side in comparison to the other are more or less stimulated depending on the angle the bee adopts in relation to the vertical (Fig. 5.3). Such groups of sense organs are also situated at leg joints and contribute to the perception of gravity [161–163] (Fig. 5.4).

The most important sense organs work together in communication. Transferring the visual angular direction determined in flight to that related to gravity on vertically oriented combs in the dark hive is remarkable. This ability is not specific to bees but found in many other insects [222]. During dance communication, directional perception in the two sensory modalities is exchanged back and forth. The dancer depicts an angular direction derived from her flight and followers use directional information from the dance on the comb in relation to gravity in finding the search area (see Fig. 1.1—green sector).

Gould exploited this close link between the visual and gravitational senses in his experiments in the following way: If a small source of light is brought into the dark hive, bees detect both gravity and light. This results in the reference point for angular direction to be a combination of both sensory inputs and the consequent orientation.

Fig. 5.3 Cushions of fine sensory hairs occur at all joints between the thorax and head, legs, and abdomen. Gravity can pull the head and the abdomen downwards as shown here (circles). The sense organs detect the deflection of these body parts in relation to the thorax, anchored by the legs to the substrate. In this way, bees can determine their orientation on the comb in the dark hive

It is not the Bees that Lie

An experimental ruse employed by Gould was to exploit "lying" bees ("We figured out a way to get bees to lie about where they'd been"; [9]) or so it was said whenever this work was mentioned later. Although irrelevant for the following discussion, studying the work of Gould, it is clear that there were no "lying bees" in this experiment. Instead, the circumstances contrived that a bee following a dance received false information (i.e. misled).

What lay behind this? The principle idea behind the experiment was the following: If one could arrange a situation in which the followers received false information from the dancers, the recruits should then use this and fly in the wrong direction. One could then see if the recruits flew to the site from which the foragers really came and returned to, and so did not use the information provided in the dance. Or they would follow the false information and fly to a site different to that from which the foragers came, in which case they used the information in the dance.

The experiment should at least answer the question of whether or not the dance influences the flights of the followers. It should establish if Wenner's belief that the dance provides no information about the location of the food source is correct or not.

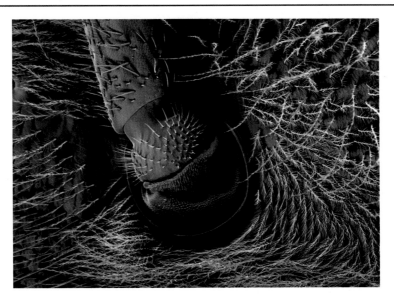

Fig. 5.4 The antennae of bees are densely packed with sensory hairs, most of which are used for the detection of odours. In addition, they detect environmental modalities such as temperature, humidity, and carbon dioxide concentration. Cushions of fine sensory hairs at joints, shown here between the antenna and the head, determine the attitude of the antenna and thereby the direction of gravity

Gould prepared two different groups of bees for his experiment. A relatively small number of bees provided dance information only in relation to gravity, and so delivered their message without falsification (they did not "lie"; for a study on bees that do "lie," see the following chapter). In addition, a second group of more than a thousand foragers that oriented themselves also to a light source installed within the hive. These combined the directional information from light and gravity and followers had, therefore, a different reference direction from that indicated by the dancers. Hence, they incorrectly interpreted the correct instructions of the dancers.

How Gould Manipulated the Dancers

Honeybees possess, in addition to the large compound eyes, three small single-lens eyes; the ocelli (see Fig. 5.1). Ocelli sense the intensity of light. If the ocelli of honeybees are occluded, the intensity of light detected by these insects with their compound eyes is diminished by a factor of six. Optical stimuli that release certain behaviours have to be six times brighter in bees with blinded ocelli to be effective. In spite of this massive limitation, such bees can still fly out of the hive and return to dance on the comb [153].

Gould managed to paint over the ocelli, embedded among the head hairs (Fig. 5.1), of at least thirty dancers (it must have been more, just how many is not evident from his study) and with such precision that a limited but defined amount of light reached the ocelli of manipulated bees. The requirement for this experiment was that the bees with covered ocelli would ignore the artificial light in the hive as an orientation cue, although the large compound eyes were intact and could see the lamp. These bees flew out of the hive and, Gould concluded from the direction they took, ignored the influence of the artificial light in the hive.

The downward force of gravity in the dark hive is the reference for the direction of the dance. Outside the hive, this is represented by the sun's azimuth (see Fig. 3.6; the azimuth is the direction to that point on the earth's surface above which the sun is vertical). The manipulated dancers in Gould's experiment directed their dances as usual only in relation to gravity and ignored the lamps. The untreated dance followers, on the other hand, influenced by the artificial light source would use this as a reference point additional to gravity. These bees should interpret the danced angles differently from that indicated by the dancers. This was the central concept of this experiment.

The difficulty of manipulating the dancers to the right extent in this experiment is evident from later studies of Maximilian Renner and Thomas Heinzeller [205]. These authors showed that only a small percentage of foragers with completely occluded ocelli returned after flying out of the hive (in one case only three out of sixteen treated bees) and these died after at most three days later. Based on their studies, they come to the following conclusion in relation to Gould's experiments:

> In the case of the ocelli-painted foragers which Gould (1975) used elegantly for his distance and direction experiments, we suggest that the covering was incomplete enough to allow further flight, but sufficient to reduce sensitivity to the simulated sunlight from the lamp in the hive. [205, p. 227]

Further:

> But a more important – indeed critical – factor is likely to be the need for checking the ocellar covering after the experiment. Even though the paint over the eye appears to be continuous at first sight, closer examination may reveal flaws. For example, when the paint dries it may rise away from the cuticle of the head, or curl up at the edges, small flakes may break away or be brushed off by the bee's cleaning movements. Hairs can make small holes in the covering, and some ocellar shields have proved to be translucent when held against a light, even though they appear quite opaque by reflected light. Only when animals caught at the feeding site can be shown by follow-up tests to have effectively blinded ocelli, can one be certain that their ocelli were blinded when they flew to the site. [205, p. 225]

These considerations are important for the Gould study because there the need was not to completely blind the ocelli like Renner and Heinzeller, but to apply a layer of paint so precisely as to allow only a defined amount of light to pass through.

This extremely difficult technical preparation of the dancers mastered by Gould could not be repeated.

Gould employed ten to fifteen foragers in each series in his experiment, in which all three ocelli were painted over as described. These manipulated bees were trained to a scented food site for which they danced in the hive. None of the other foragers, some of which turned up as dance followers, were manipulated.

Gould Repeats von Frisch's Experiments—With a Refinement

Two experiments are presented here with which Gould aimed to come as close as possible to the fan studies of Karl von Frisch.

Six control stations (equivalent to the scented plates of von Frisch) were arranged in a fan, 400 m away from the hive with an angular separation of three degrees from each other in relation to the hive. The control stations consisted of a scented trap from which the bees could not escape (Details of the setup of the stations in [95]). The feeder was located a few metres behind the control station at the centre of the fan (Fig. 5.5) or behind the control station at the eastern outer edge of the fan (Fig. 5.6).

In both experiments, fifteen marked bees with occluded ocelli, in the 50-min period of the study, flew back and forth from the hive to the feeder and danced in the hive. Recruits trapped in the control stations during the experimental period were counted.

Figure 5.5 shows that most recruits arrived at the control station directly in front of the feeder continually visited by foragers. During an observation period of fifty minutes, the distribution of recruits corresponded principally with von Frisch's classic fan experiment (Fig. 3.5). Again, this study does not allow an estimate of how important the dance was for the recruits. 88 recruits came directly to the feeder with the experienced bees, and 77 reached the control station directly in front of it.

Recruit distribution could be interpreted to have occurred without assuming a role for the dance: The recruits could, according to Wenner, have oriented to an "odour concentration."

This critical objection no longer applies to Gould's experiment 10, described below.

In experiment 10 (Fig. 5.6), most recruits, 82, also flew directly to the feeder visited by the dancers during the fifty-minute observation period. 68 recruits arrived at the control station that lay directly in front of the feeder and over which the trained foragers flew. However, 30 recruits arrived at a station indicated by the light in the hive. They had interpreted the dance according to this reference and taken the wrong direction.

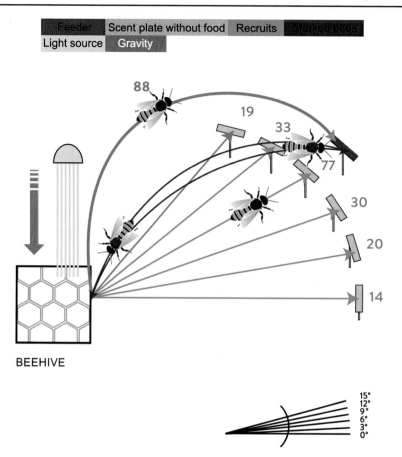

Fig. 5.5 The figure represents Gould's experiment 9 [95]. He trained fifteen bees to a feeder (red flight path) that danced in the hive to recruit others. An artificial light source was installed in the hive so that the direction of the light (yellow lamp) was the same as the gravitational reference point (grey arrow). Directional information was the same for both the dancers and the dance followers. The majority of recruits arrived at the feeder or at the control station directly in front of it (number and flight paths of the recruits shown in blue). The angular separation between the stations in the figure is exaggerated; the separation between the outermost stations in Gould's experiment was fifteen degrees (see the sketch at the bottom right)

Experiment 10 and others in Gould's dissertation were the first in research on bees to clearly demonstrate that the bee dance influenced, without other signals and cues, the area in which recruits searched for scented goals. In more than half of the experiments, the majority of the recruits arrived at control stations and not at the feeder.

The influence of the dance on the distribution of the recruits in the field was even clearer in fan experiments where Gould spread the control stations and the feeder over an angle of 90° and repeated the experiments described above (Figs. 5.5 and 5.6).

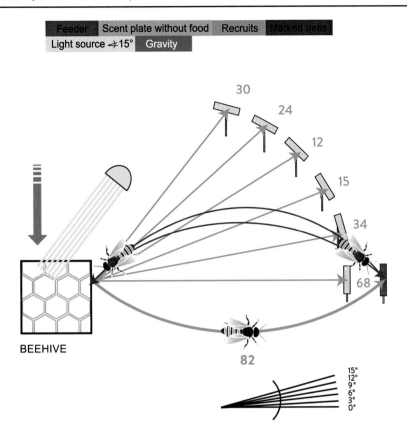

Fig. 5.6 Gould's experiment 20 is shown here [95]. As before, Gould trained manipulated foragers to a feeder (red flight paths) and installed an artificial light in the hive that now illuminated the comb on which the dances took place from the side. The direction of the light diverged fifteen degrees from the vertically oriented gravitational force. The reference direction indicated by the artificial light (yellow lamp) is shown schematically in the figure and to which the follower bees were sent, namely the control station at the extreme left of the fan. On the other hand, dancers using gravity (grey arrow) as the reference indicated the feeder to which they had been trained and visited (marked orange) and located behind the control station at the extreme right of the fan. Here too, most of the bees arrived at the feeder, although a significant number of bees, 30 in all, flew to the control station indicated by the artificial light source (flight paths and the number of bees in blue). The angular differences are exaggerated in the figure; the angular difference in Gould's study between the extreme left and right control stations was fifteen degrees

Gould also conducted step experiments in which the distribution of recruits over the control stations showed they were not only influenced by the direction given in the dance. Many recruits in the step experiments arrived at or near control stations that coincided with the distance indicated in the false dances, although not visited by the dancers.

In the step and fan experiments of Karl von Frisch, recruits that arrived at control stations could also have either lost contact with the experienced bees in the field or never found it and followed the scent to a control station. There was no need to explain their behaviour by information about the locality gained from a dance. In contrast, Gould's experiments showed that many recruits arrived at control stations, the direction and distances of which were indicated in the false dances, although not visited by the dancers.

While both Gould and von Frisch scented all the control stations in their experiments, it is not possible to differentiate and compare the three phases of distant goal orientation (SEND, SEARCH, and ATTRACT). The independence of the first stage of distant goal orientation was nevertheless, for the first time in bee research, demonstrated by these experiments. The dance alone sends the dance followers off on flight paths that, within a certain area, are independent of the flights of experienced bees.

The data from Gould's experiments also show the massive influence foragers flying in the field have on the recruits. Most recruits were caught directly at the feeder, the site visited by the experienced foragers; a result obtained by von Frisch as well (see Fig. 3.4). Many recruits also arrived at the control station in front of the feeder (see Figs. 5.5 and 5.6).

Gould summarised his work in the following cautious conclusion:

> Depending upon conditions, honey bee recruits use either the dance language AND odor information, or odors alone. Wenner's conclusion that "one cannot have it both ways" – that bees have only one strategy for recruitment which they must use under all circumstances – is clearly incorrect. On the other hand, recruitment to odors alone might be the usual system in honey bee colonies not under stress. [96, p. 691, emphasis in the original]

Gould did not take communication between bees in the field into account. In his careful conclusion, he suggests orientation to odour as the normal procedure. He excludes the discovery of a goal through information in the dance alone and without additional orientation cues (see Fig. 5.7).

William Towne Confirms the Thesis of Karl von Frisch

It may surprise that Gould's conclusion from his data was taken by the research world as the final confirmation of von Frisch's model for recruitment [117, 178].

Step and fan experiments were the subject of another doctoral study ten years later by the American William Towne [265]. Towne's experiments were laid out like those of von Frisch except with a larger number of control stations in order to get a better spatial resolution of the distribution of recruits in the field.

Individual marked bees were trained to a feeder located in the centre of a broad arrangement of control stations. Twenty experienced bees flew back and forth between the hive and centrally situated feeder and brought dance followers to fly out of the hive. Towne recorded recruits captured at the 25 control stations.

His results confirmed those von Frisch had obtained from his own step and fan experiments.

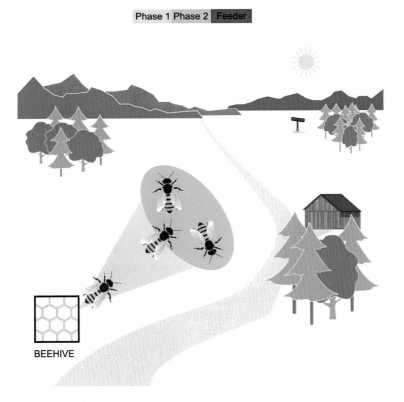

Fig. 5.7 In his experiments, Gould could show that the dance influences the direction and the distance a bee flies to search for a feeder (marked orange), not, however, that the recruits can find the site from the instructions in the dance alone. This demonstrated that the dance determines Phase 1 of distance orientation (yellow sector: SEND), where Phase 2 (green field: SEARCH) begins (see Fig. 1.1)

The closer the control station lay to the centrally located feeder, the more recruits were caught there. Towne, like von Frisch before him, conceded that in his analysis, he had not taken the behaviour of dancers or recruits arriving at the feeder into account.

Summary: The Dance Sends Bees to An Area—No More (and No Less)

If we take a different point of view and give the importance of communication between bees in the field that von Frisch originally assigned to it [72], the impression differs from the more recent authors of step and fan experiments. The distribution of bees over the control stations appears to be more the result of an interruption in the signal chain between the hive and the goal.

Gould interpreted his results in the light of the controversy as to whether or not recruits use the information from the dance. His answer to this question was as follows: The dance information influences the behaviour of recruits flying from the hive. It is, therefore, no surprise that Gould also did not take communication in the field into account, despite his observations that a large number of recruits (in Figs. 5.5 and 5.6; see Tables 1 and 2 in [96], there the majority) flew to where the dancers from the hive were also underway.

Gould did not comment on the behaviour of the experienced bees at the feeder. Although many ("a substantial number"; [96, p. 691]) recruits arrived at the feeder (what is meant by "a substantial number?"), it did not occur to Gould that the experienced bees could have assisted the orientation of the bees in the field.

What remained after Gould's careful interpretation of his own data? At the least, he determined that Adrian Wenner was incorrect in his opinion that the dance has no influence on the direction taken by recruits on their flight from the hive.

All fifteen of Gould's experiments showed that the dance provided information about both direction and distance. The distribution of recruits over the control stations clearly demonstrated that experimental manipulation of directional information received by recruits about the location of the site influenced them.

However, these experiments do not mean that precise goal orientation can be derived from the dance alone, as implied in the classical formulation:

> After they have observed many dances and followed the dancer, the dance followers are in the position of being able to find the food source on their own. [168, p. 58]

In relation to the three phases of distant goal orientation, the results of Gould's work are limited to the first phase. The dance does not send recruits just anywhere in the field but to a sector and distance from the hive where support from the dance ends.

Research Focuses on the Inside of the Hive

6

The idea of a dance language directed the attention of research almost exclusively to studying the interaction between transmitters and receivers in the bee dance in which the secret of the consequently precise recruitment must be hidden.

Events related to the recruitment of foragers to a site in the field are divided into three sequential occurrences: Those in the hive, those at the site, and what happens on the stretch between the hive and the site. Three activity phases between the start and goal constitute distant goal orientation (see Fig. 1.1).

Studies of recruitment behaviour in the field were limited, until recently, to counting bees at artificially erected feeders and surrounding control stations visited by experienced bees and to measuring the flight times of bees between the hive and feeder. Only a few studies took up the role of experienced bees at the feeder.

Since a few years, radar techniques have enabled the flight paths of bees to be tracked through the wide field, in its truest sense, between the hive and the goal of the forager bees (Figs. 3.1; 9.2 and 9.3; see also [208–210]).

Gould's experiments confirmed the work of von Frisch: The dance ensures that recruits leave the hive with an initial perception and they do not distribute themselves randomly in the field. Because of von Frisch's discovery, the main interest of bee research for the next decades focused on increasingly deeper analyses of the bee dance. The aim of the research program was to investigate every imaginable and measurable aspect of the interaction between dancer and dance follower in the hive and to analyse its suitability for the precise site description.

The hypothesis: The dancer as a sender transmits information about the location of a goal to her dance follower, the receiver. The information that will lead the follower, to, or more carefully expressed in the proximity of, the goal, must reside in the interaction between the dancer and the dance follower.

Technical advances in research methods opened new windows and brought new details to light. Slow motion videos allowed an accurate description of dance movements, and recordings of the dancer's body movements led to an exact spatio-temporal description of the dance (see Fig. 7.1). Physical and chemical

© The Author(s), under exclusive license to Springer Nature Switzerland AG 2022   57
J. Tautz, *Communication Between Honeybees*,
https://doi.org/10.1007/978-3-030-99484-6_6

measurements enabled airborne sound, movement, comb vibration, chemical signals, and electrical fields that accompanied the dance to be recorded (see Chap. 7).

Round and Waggle Dances Say Something About the Distance and Quality of a Food Source

Two hundred years ago, Unhoch accurately described the dance:

> A single bee […] trembles her raised abdomen for a short period and the neighbouring bees do the same, press their heads down onto the substrate, finally all turning together through a little more than a semicircle, to the right and then to the left up to six times, in a formal round dance. [270, see p. 115]

With the passage of time, the behaviour of the dancer and follower was dissected in increasing detail. How does the dancer move and how do interested follower bees respond?

Karl von Frisch published the first detailed account of the waggle dance:

> A bee […] crawls up the comb and begins to turn in the middle of other bees; but not through a full circle as in a round dance, instead initially a semicircle, then walks in a straight line over 2 or 3 cells back to the start, turns now to the other side and moves along a second semicircle to close a full circle with the first, walks again in a straight line along the diameter of the circle to the start, then repeats the first semicircle, walks again to the start, repeats the second semicircle and so on so that the semi-circular paths alternate with the straight path to the starting point. Movement along the semi-circular paths is normal whereas movement along the straight path is always accompanied by waggling. This consists of rapid, rhythmic movements of the entire body from side to side […]. [72, p. 73]

Von Frisch distinguished the waggle dance from the round dance, another dance form, because he noticed the dance figures could differ from one another. Initially, he attributed the round dances to the nectar collectors and the waggle dances to the pollen collectors [74]. During the research of the dance behaviour in his step experiments, he realised the connection between the dance figure and the distance from the hive. The round and waggle dances he had previously described were now correctly assigned: Round dances advertise food sites that are located near the hive; they contain no elements related to distance. Waggle dances advertise distant goals and contain elements correlated with distance (Fig. 6.1). However, he did not determine any sharp distinction between the two.

As an intermediate von Frisch described the so-called sickle dance (see also Fig. 7.4) and established through his simple direct observation what research later confirmed with highly sophisticated technical methods: There is no clear boundary between round and waggle dances. The dance gradually changes its form with increasing distance between the hive and the food site; the proposed categories (round and waggle dance) have no sharply defined boundaries corresponding to reality [90]. All dances include elements of the waggle dance. Directional elements are used even in dances for sites at the shortest distances from the hive [90, 127, 138].

Fig. 6.1 Three phases of movement in the waggle dance taken from an initial sketch of von Frisch's: Return run to the left, waggle phase, return run to the right. A complete dance cycle lasts only a few seconds and takes place in the space of about two to four centimetres in diameter. The representation follows the original figure from [75, p. 11, Fig. 6]

The mistaken idea of two clearly separated categories of "round dance" and "waggle" dance nevertheless persisted. Repeatedly in the literature, a sharp boundary between the two is given (about 100 m, varying slightly depending on the race and species; see also p. 135).

The dance does not only contain information about the site locality. Karl von Frisch had already noticed that forager's dances promoting similar feeders distinguished between the goals according to their worth. It appeared to him that dances were "livelier" the more attractive the food was for the forager. The "value" of a goal depends on the cost-effectiveness of foraging at this goal, determined by factors such as sugar concentration, the abundance of sites, and distance to the source [39, 54, 186, 187, 272, 286].

A modern analysis of dance movements has explained where the impression of varying "liveliness" comes from. The speed of the return phase which, as far as is known, conveys no site information can vary and is correlated with the attractiveness of the food site [121, 235]. The more attractive a site is for a forager, the more rapidly she runs on the return phase of her dance back to the start, and the livelier the dance appears to an observer. In addition, irregularly distributed vibrational pulses produced by the wing muscles occur during the return phases ([157]; Fig. 6.3, for details see p. 98ff.), the basic frequency of which lies higher for more attractive food sites. The performed dances consist of many rounds, up to one hundred [234] or even more depending on the attractiveness of the food source [18, 53, 78, 228, 229, 234, 235]. The same is true for round dances. Their total duration and number of turns increase with the rising concentration of sugar in the food [272, 273, 274].

True for all forms of the dance is the quality/attractiveness of the food source; the higher, the more likely dances will be performed [61, 121, 228, 234, 273]. The behaviour of bees at the completion of their outward foraging flight, including their communication behaviour at the goal, is also influenced by the attractiveness of the food site. This in turn is dependent above all on the nature of the food and the difficulties confronting foragers on their flight to the site. Josue A. Nunez (1924–2014) was a pioneer in the research on these fascinating inter-dependencies.

Slow Motion Recordings Reveal the Waggle Dance Movements

A dance round lasts for only a few seconds and the details cannot be captured by an unaided human eye. The first detailed view of the complex movements of the dance was made possible with high-speed video recordings.

These showed the following: During the waggle dance, the bee does not walk forward on her six legs with the usual pattern of a six-legged insect. Actually, she does not walk at all. The behaviour is better described as a waggle stand, because each leg clings to the substrate for as long as possible while the side to side rocking body leans forwards over the stationary legs. Only when the forward movement of the body has reached its limit and the legs cannot stretch any further, each leg takes a small step, independently and not co-ordinated with another leg, a walking movement unknown in any other insect. Following a suggestion [261], the term "waggle run" has been replaced in the more recent literature by the more appropriate term "waggle phase."

This new view and terminology have the additional advantage that the motion can be identified in dances differentiated in the old system as round dances (in that according to the older ideas and definition were not waggle dances. For the functional consequences of these unusual forward movements, see p. 62ff).

The Follower's Ballet—A Few Insights

For an observer, the followers appear as single, or groups of bees, because they move with the turns of the dancers. Followers position themselves mostly to the side or behind the dancer in order to remain in contact with the dancer as long as possible. The best way to achieve this is to place themselves in front of the dancers at the end of the waggle phase and change to the other side as soon as she begins her return run. As a rule, the followers attend between five and eight dance rounds before they leave on a foraging flight [56, 131, 166, 226].

Details of the follower's movements are as stereotyped as those of dancers but have not, with a few exceptions, attracted the same interest of researchers. Minimally, one knows that the extended antennae of the followers are held at a particular angle, in order to maintain the best possible contact with the dancer. During the turns in the ballet with the dancer on the vertical combs, the abdomens of the followers are not moved in the same way as when the bees turn normally. Instead, they make abrupt saccade-like movements alternating with pauses. The saccades pull the abdomen, if it has been deflected by gravity, into line with the thorax [218].

One can speculate that these saccade-like movements of the abdomen during the turns of the dance are relevant for establishing the direction of the vertical gravitational force mediated by the output from the sensory hair cushions between the thorax and abdomen (see Fig. 5.3) that are only in equilibrium when the thorax and abdomen are aligned [218].

Comb Vibrations Attract Dance Followers

The highly complex behavioural chain in forager recruitment begins in the dark hive when dancers and dance followers meet. What signals does the dancer use to attract attention to herself?

Every active movement of an animal generates associated physical forces and fields, measurable in their vicinity. Dancing honeybees produce air movement and airborne sound caused by wing oscillation, especially during particular phases of the waggle movements. Waggle movements and wing oscillations are transferred as vibration to the comb (Fig. 6.2). In addition, the dancer's body temperature is higher than most bees around her. Activity driven by muscle contraction in the entire animal kingdom, including dancing bees, induces electrical fields. Dancing bees also emit a specific odour.

That physico-chemical phenomena can be measured does not mean that they are meaningful for the behaviour of animals. Only observation of the animal's behaviour can determine this.

Modern recording methods significantly aid in gaining better insights into behavioural details. High-speed videos are an important technique. Recordings of dancers and followers during a waggle dance allow microbehaviour of the participants to be analysed in detail. It is even possible to let time run backwards by reversing the recorded videos. By such means, one can determine from which direction and distance a follower was attracted and her behavioural reaction when she first detects the dancer.

All of the above-mentioned signals could be candidates should the dancers and the followers be standing on the surface of the same comb, which is the usual situation. However, vibrations of the comb from the dancer will not reach followers on separate combs. Only those signals not mediated by the comb can spread to another bee; however, these do not attract followers. Without comb vibration, dancers cannot attract followers from a distance [258] (Fig. 6.2).

On rare occasions, foragers on the surface of a comb directly opposed to that on which the dancer is active can be observed to follow the dance, although the participants are back to back across the space between the combs. Such followers are not attracted from a distance as those on the same comb surface as the dancer. They also do not turn their heads towards the dancer and move towards her from a distance as do those on the same comb as the dancer. Back to back followers only dance consistently with the dancer if they happen to touch the dancer with their antennae and identify her through direct contact [258].

The observations described here indicate comb vibration to be the dominant signal in leading dancers and followers to one another.

Fig. 6.2 The foot of a bee. The dancer generates vibration signals and transmits them with her six feet to the comb, detected by dance followers—also with their feet on the comb!

A Telephone "Landline" Unites the Partners

The upper rims of comb cells are thickened into a bulge. These play a decisive role in communication between bees in the darkness of the hive and in particular in bringing dancers and dance followers together (for details see [256]).

Karl von Frisch speculated that faint vibrations could play an important role for bees in the dance language. A simple behavioural experiment showed that his suspicion was justified. If bees are allowed to dance on empty combs, the indicated feeder is visited by three to four times the number of recruits in comparison to when the dance took place on combs with closed cells. Dances on empty cells attract more attention in the hive, are attended by more followers, and recruit more bees to the food source [255].

The physical reason for the different effectiveness of identical dances, but performed on different substrates, was revealed by employing a highly sensitive displacement measuring technique, laser Doppler vibrometry. This device allows remote detection of unbelievably small vibrations that a dancer induces in the comb.

The thickened cell rims of all the cells together form a six-sided mesh network. This rests on the thin walls of the cells that allow it to move easily over small distances in the plane of the network, like the net that hangs around a soccer goal.

Small displacements spread along the thickened cell rims over the entire surface of the comb. Physically, these are neither longitudinal nor latitudinal but instead a rapidly dispersing deformation. The net optimally conducts a narrow frequency band between 230 and 270 Hz (cycles per second) as a *comb-wide web* (a network of six-sided meshes formed from all cell rims of a comb). Displacements in this frequency range are amplified; the cell rims move through the greatest possible distance. Remarkably, whether the cells are empty or filled with honey has no effect. Only cells sealed over with wax stop the transmission. If a dancer stands on closed cells, vibrations cannot be detected in empty cells neighbouring a sealed comb area. Should an island of sealed cells occur in an otherwise unsealed area of cells, the vibrations pass around the sealed area [217].

The narrow frequency band of 230–270 Hz, in which the comb optimally conducts mechanical vibration, corresponds with the frequency of short vibration pulses that dancers emit during the waggle phase of their waggle dances (see Fig. 6.3). Honeybees control the construction of their combs down to the last detail and tune their telephone network to conduct their own communication frequency best. Material properties, architectural design, and bee behaviour are all perfectly adapted to one another.

Empty combs and those not occupied by bees conduct the faintest vibrations along the cell rims across the entire network. Combs occupied by a number of bees large enough to load the comb with their weight act as though the upper surface was sealed. The vibration is damped, spreads only a few centimetres, and reaches bees only up to this distance away. The area included in the range reached by the vibratory message of the dance is, therefore, limited to what is practical in a communication biological sense.

How Do Bees Extract Weak Signals from High Background Noise?

Communication signals normally exceed background disturbance. In the world of sound and vibration, this means the signals are louder and more powerful than the disturbance in the background. This does not apply to the waggle dance vibrations. Several thousand bees on the same comb, all active and occupied with various tasks, generate a continuous noise from which the weak communication vibrations do not emerge as strong signals. How can such weak signals be recognised?

Astronomers solve the detection of weak signals in high background noise by coupling two widely separated antennae. Comparison of the signals they receive allows weak regular events from distant radio stars to be identified.

In principle, honeybees are similarly equipped. Each bee possesses with its feet six separate contacts with the cell rim network. Vibrations are detected by all six feet and compared, like radio astronomers in their observation of stars.

A comparison of measured vibrations from several different points on the cell network of a comb allows one to identify a pattern, not observable at a single point, which can be resolved despite the strong background noise. Vibrations that spread across the net as displacements of the cell rims form a remarkable flat, regular representation of cell rim movement.

The simplest case, an induced vibration of a single cell rim, leads to the following: The opposite cell walls of a neighbouring row of cells move in synchrony and in the same direction. However, in just one cell in the row, the opposite walls move in opposite directions. The material properties of the wax and geometry of the cells and complete comb determine this complex movement pattern [260]. Because the dancer pulls with all six legs on the rims of the cells, it would be expected that around the dancer, as the transmitter of vibration, many such pulsing cells would occur. A follower, the receiver of the signals, also stands on the cell rims and spans up to three cell widths with her legs. With the vibration-sensitive receptors in her legs, she could detect the two-dimensional displacement pattern in the dark without difficulty.

Behavioural analyses of video recordings support this view. Tracing a follower that attended a number of dance rounds, in video recordings run in reverse, back to the beginning of the dance ballet and further, allows the location on the comb to be determined where the follower became aware of the dancer. If she detects the presence and the direction of an active dancer, she turns her head towards the dancer. Thereafter, she turns her body towards the dancer and runs in the appropriate direction until she bumps into the dancer and immediately participates in the ballet. Superimposing the position of the "pulsing" cells, determined from the physical measurements, and the position where the follower detected the dancer, determined from the behavioural results, reveals that they correspond. The "pulsing cells" and the "I detect a dancer" locations are the same.

These observations suggest it is highly likely that the two-dimensional pattern of comb vibration leads the bees to a dancer even over an extremely noisy comb. Dances performed on immovable surfaces or on the bodies of other bees, such as in a swarm cluster, do not attract followers from a distance.

A priority in communication research in honey bees, based on the importance assigned to the dance, has consequently been the search for signals the dancer transmits. These signals, relevant in the direct contact between dancer and follower, could contain the information that leads the recruits to the advertised goal.

The Difference Between Signals and Cues

Behavioural researchers draw a distinction between signals and cues. A simple definition could be the following: A signal is everything that, from the sender's perspective, "intentionally" alters the behaviour of a receiver a cue, is everything that "unintentionally" changes the behaviour of a receiver [167].

A signal mediates information to a receiver, about the sender or its surroundings, and is in most cases advantageous for both sender and receiver. A cue is only advantageous for the receiver. The position of the sun in the sky is a cue from which the bee can derive directional information. For the sun it is irrelevant, but not for the bee that can use the information for orientation.

Although the explanation may appear simple, individual examples can be complex and interesting. How do communication biologists categorise the flower scent adhering to a returning forager, detected by the hive colleagues, and their consequent foraging flights to the same species of flowers? The mediator of the scent is the returning forager to which the scent adheres [84]. Is the scent a signal or a cue? Who profits from the behaviour of the bee released by the scent of the flower? Certainly, the flowering plants are pollinated by the visiting bees. The scent is, therefore, a signal of the flowers. However, what about the bees in the hive that detect the scent of their nestmate? They are also advantaged because all bees in the colony profit from successful foraging activity. Then the flower scent would be signals for both flowers and bees.

Signals are formed and their use is adapted through selection. They often emerge from cues that are initially meaningless to the sender [164].

The bee dance has a clear communicational function and during recruitment behaviour provides signals between bees about the nature of the goal (flower scent), the attractiveness of the goal (livelihood of the dance), and assistance to find a search area from which the goal could be found (waggle dance).

The conduction pathways are decisive for the transmission of signals. They determine if and how the signal will be effective.

For bee dances on the comb (and on the bodies of other bees in a swarm cluster, see Fig. 11.2), the substrate on which the sender and the receiver stand and the air between them come into question as conduction pathways.

The effect dancers have during their dance behaviour on their surroundings and how far these came into question as communication signals were measured and analysed with experimental methods, technically the most advanced prevailing at the time [173].

"Deaf" Bees Are Highly Sensitive to Vibration

Historically, the sound generated by the waggle phase was the first to be discovered and offered the clearest criterion of the duration of the waggle time [53, 58, 113, 176, 177, 269, 282]. The sound is produced by vibration pulses of bee flight muscles [119, 120, 156, 241, 255]. This generator is the honeybees' most powerful performer. When not flying, she can partially uncouple the wings so that they move only slightly when producing sound. As a result, a so-called near-field sound effect occurs in the direct vicinity of the dancer; a typical three-dimensional structure that other bees can use as a source of information [136, 173, 176].

Honeybees are deaf; they have no sense organs for the perception of sound pressure waves. However, if the air displacement of a sound pressure wave is large enough, as in near-field sound, bees could detect this with the sensory hairs on their heads or with the Johnston organs in the antennae (Fig. 5.1) [46, 119, 134, 137, 268, 269].

In addition to near-field sound, the wing movements produce directed air currents that can stimulate the same sense organs on the follower's heads that respond to the near-field sound [173].

Sound production takes the form of short pulses, occurring at regular intervals during the waggle phase. Dancers produce a vibration pulse at each reversal of the back and forth movement of the waggle dance (Fig. 6.3). The pulses have a basic frequency corresponding to the wingbeat frequency of 230 to 270 beats per second and a duration of about fifty milliseconds. Such pulses can sometimes be observed during the return run, but appear irregularly [157].

Fig. 6.3 An idealised waggle dance figure to clarify the occurrence of embedded vibration pulses. These are shown as blue oscillations under the magnifying glass and represent high-frequency pulses accompanied by vibration, sound, and electrical field events. The pulses occur during the waggle phase at each reversal of the body movement [52, 260] and irregularly in the return phase [157]

The "deaf" bees are far less sensitive to airborne stimuli than to faint substrate vibrations they perceive with their legs [123, 133, 217]. For this communication channel, bees exploit a mechanical ruse to transfer vibrations of the wing musculature through their legs into the comb, aided by the waggle dance movements that we find so remarkable. A single standing bee or even a lightweight dancer running over the cell rims would transfer very little energy through her thin legs into the comb. However, because, during the waggle dance, her feet cling to the cell rims, she can bring the cell rims to the left and right alternately under tension and this tension is greatest at the reversal of the body direction. The dancer chooses exactly this moment to inject the vibration into the comb.

The bees emphasise each reversal of body movement with a vibration pulse that can spread out over the rims of the neighbouring cells [157, 181, 217, 255, 257]. "Muted" dances that occur now and again show that vibration pulses as near-field sound and comb vibrations are important signals in the waggle dance. Mute waggle dances appear to the human observer to be "exaggerated" but the vibration pulses are absent. Mute dancers attract no followers even when performed in a crowd in an attempt to draw attention [55].

Electrical Fields Support Communication

An additional signal pathway in the bee dance, known for a long time [289] and recently investigated in detail [107], is modulated electrical fields accompanying the vibration pulses. The timing of the electrical events corresponds with the appearance of near-field sound and the vibration pulse and carries no additional information; all the signal forms are interchangeable.

Dance Followers Receive the Same Information in Parallel

Substrate vibration, nearfield sound, air currents, and dynamic electrical fields all occur at the same time in the waggle dance as the result of the activity pattern of the dancer's flight muscles. All mediate the same information and, in other words, represent redundant signals. Information is redundant, if, as the signals in the waggle dance, the same information is present in parallel. Which signal form the dance followers pick up is communication biologically irrelevant. The precision of information about the site in the dance is not improved if the dance followers pay attention to more than one signal. This fact is repeatedly overlooked. In relation to the clear imprecision of the dances of bees, one frequently comes across the idea that through different parallel signals, the exactitude of the information would improve. This is not the case.

Redundancy improves the susceptibility of the signals to noise, a considerable advantage in the crowded tumult and continuous activity in a bee colony. The synchrony of all the identified signals in the bee dance raises the certainty that the message will arrive despite disturbances or the failure of one or other channels.

Chemical signals from the dancer cannot carry any spatio-temporal structural signals like those produced by the flight musculature, but would help her to be found on the crowded comb [22, 93, 263].

The Position of Dance Followers Is not Critical

Dancers and followers are close during the dance, and in the waggle phase, lateral swings of the dancer's abdomen bring it into contact with the dance follower's antennae. The regular waggle movement leads to a regular tactile pattern received by the antennae of followers. In principle, the antennal mechanoreceptors, responsive to mechanical stimuli, would reveal the attitude of the dancer on the comb and the duration of the dance. Because the dance follower knows her position relative to gravity from her own gravity receptors (Fig. 5.3), in principle she could determine the approximate angle relative to vertical on the comb along which the dancer waggles [91, 213].

Dance followers do not distribute themselves evenly around a dancer, but instead they position themselves mainly to the side or behind her [21, 131]. The accuracy of forager flights is independent of the follower's position relative to the dancer during the dance [253]. This result came from an experiment in which three experienced foragers flew back and forth to a feeder and danced for it in the hive. It could be shown that recruits arrived at the feeder regardless of the positions they occupied in relation to the dancer during the dance following [132]. This observation also speaks for a most reliable and robust recruitment mechanism in that it does not rely on every detail of the dance communication in the hive.

Every study of the dancer's signals during the waggle phase aims to explain the precision with which the recruits arrive at the goal. Neurobiologists even searched in the brains of bees for elements associated with recruitment communication [1, 2, 3, 23, 125].

Measurement of the Dance Figures in Space and Time

7

The limits of measuring devices determine the accuracy of what can be measured. The appropriate accuracy of the chosen measurements depends on the question and the context of what is measured.

Two of von Frisch's discoveries led to taking exact measurements of the dance and every conceivable parameter connected with it: First, the changes in dance figures with the location of a goal in relation to the hive (dance language) and, second, the exact arrival of recruits at this goal (success of the dance language). Both focused research on the dance procedure, and a search for the information transfer between dancer and dance follower during the bee dance. No wonder that to this day the lion's share of research is occupied with measuring every thinkable detail of the bee dance, and because a glass observation hive affords simple access to the first stage of recruitment behaviour.

The dance indicates the direction and distance of the goal—hence one must examine every component of the dance behaviour to see which would best be suited for direction, and which for distance.

Karl von Frisch made a beginning with a protractor on the glass wall of the observation hive and a stopwatch, to study information about the site area indicated by the dance figure.

Today, Bee Dance Movements Are Measured Precisely

Two technical advances over the last few years have enabled every detail of bee dance movements to be clarified: slow motion video recordings, and automated tracking and recording of the exact course of dance paths in space and time, with bees carrying markers that are electronically or optically detected.

Before beginning an analysis of channels through which a dance follower could receive the information, it must be clarified if the dance itself is sufficiently precise and reliable to serve as the basis for a message that will lead recruits to a goal.

© The Author(s), under exclusive license to Springer Nature Switzerland AG 2022
J. Tautz, *Communication Between Honeybees*,
https://doi.org/10.1007/978-3-030-99484-6_7

The pathway taken by a dancer in the waggle dance is represented classically as a smooth sine wave with a clear orientation from which the directional and distance parameters can be derived (see Fig. 3.6).

Modern methods allow space and time recordings of a dancer's pathways on the comb surface. Examining these recordings (see Fig. 7.1), the first impression is the irregularity of the single pathways. Just where exactly the waggle phase begins and ends is not clear.

This uncertainty also applies to the indication of direction by a waggle phase. One should be able to derive both accurately if the location of the food site is sought in the field according to their correlation with it. Faced with the "noisy" dance pathways one is not much better off, despite modern methods, than the previous generation who obtained corresponding values directly from dances with a protractor and stopwatch.

Fig. 7.1 An automated optical recording of abdominal movements of a dancer with many consecutive waggle phases and return runs (after 190)

In fact, every dance path is so distorted and bent (Fig. 7.1) that a determination on the basis of direction and length of the waggle phase leads to no strict criteria for an observer. Both parameters can only be estimated.

Estimating values to find an approximation of the true value requires the range of the measures to be available. In this manner, quantitative measures over a usually normally distributed range of uncertainty limits the region in which the true value can be determined.

Measurement Errors Appear to Play no Role for Bees

The variation of individual values underlying the dance movements is of little importance because it is completely submerged in the "noise" of variable dances. Sequential rounds of the same dancer differ from one another. This applies to each new dance of the same dancer and is even greater between different dancers. The collected data cover a range that is larger than the expected measurement error. This may be why the accuracy of measurements of dance movements receives little or no mention. In only two studies, and then only from automated data collection [145, 200], can one find an estimate of the measurement errors (the accuracy) due to the scientific collection of data from the dance.

If a bee researcher knows the location of a goal as the reference point, because he, after all, had arranged it, every value taken from the dance that differs from this is registered as an error. Errors in the dance information advertising a site would have consequences for bees if they prevented dance followers in the field from finding the orientation cues leading them to the goal. Such a dance language error does not occur even with a variation of forty degrees of arc—the recruits still arrive at the goal (see misdirection p. 74ff). Recruitment is robustly protected against the inexact information in the dance.

VON FRISCH'S Detailed Measurements of Dance Figures

Karl von Frisch made the astute observation that for distant food sites, the waggle dance indicated a direction and that this changed systematically with the sun's position. He wanted to know exactly and consequently measured the direction of dances with a protractor held against the glass walls of his observation hive. He and his co-workers determined the direction of the waggle phase in relation to the vertical (i.e. gravity) and recorded angles to an accuracy within 0.5 degrees of arc (Fig. 7.2).

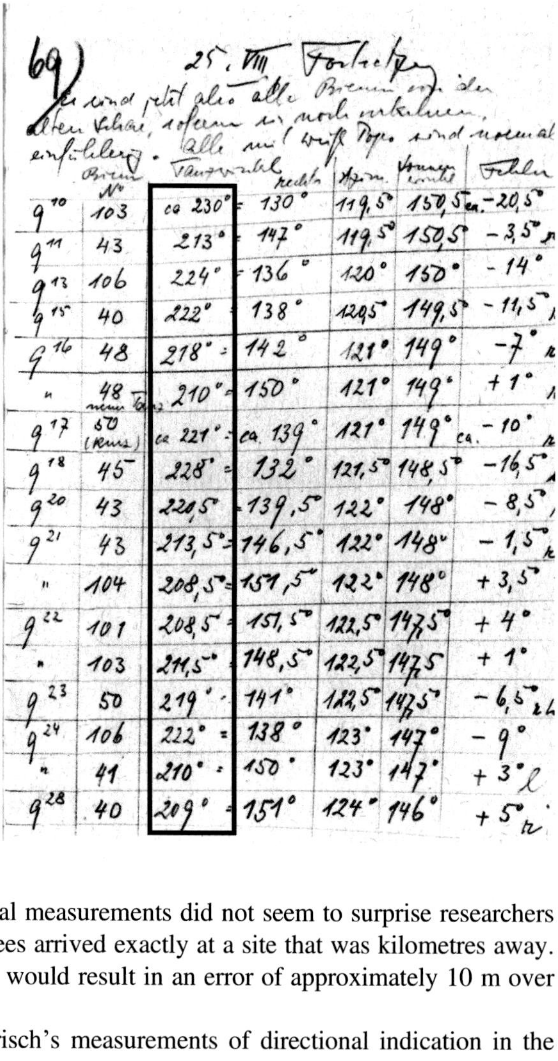

	Biene Nr.	Tanzwinkel	rechts	Azim.		Fehler
9^{10}	103	ca 230°	130°	119,5°	150,5° ca.	-20,5°
9^{11}	43	213°	147°	119,5°	150,5°	-3,5°
9^{13}	106	224°	136°	120°	150°	-14°
9^{15}	40	222°	138°	120,5°	149,5°	-11,5°
9^{16}	48	218°	142°	121°	149°	-7°
"	48	210°	150°	121°	149°	+1°
9^{17}	50	ca 221°	ca. 139°	121°	149° ca.	-10°
9^{18}	45	228°	132°	121,5°	148,5°	-16,5°
9^{20}	43	220,5°	139,5°	122°	148°	-8,5°
9^{21}	43	213,5°	146,5°	122°	148°	-1,5°
"	104	208,5°	151,5°	122°	148°	+3,5°
9^{22}	101	208,5°	151,5°	122,5°	147,5°	+4°
"	103	211,5°	148,5°	122,5°	147,5°	+1°
9^{23}	50	219°	141°	122,5°	147,5°	-6,5°
9^{24}	106	222°	138°	123°	147°	-9°
"	41	210°	150°	123°	147°	+3°
9^{28}	40	209°	151°	124°	146°	+5°

Fig. 7.2 The original protocol from the work of Karl von Frisch in which he and his co-workers measured the angular direction of waggle phases in relation to gravity. They recorded the angles to an accuracy of 0.5 degrees of arc

Such accuracy of individual measurements did not seem to surprise researchers because, after all, recruited bees arrived exactly at a site that was kilometres away. A discrepancy of 0.5 degrees would result in an error of approximately 10 m over 1000 m.

The result of Karl von Frisch's measurements of directional indication in the dance was a classic illustration that must be one of the best-known representations of animal behaviour (Fig. 3.6). It has been reproduced countless times, more or less in its original form, to this day.

The illustration asserts: A forager provides the direction to a food site in her waggle dance. In the field, this direction relates to the sun' position (more accurately, the sun's azimuth). Within the dark hive, the downward directed force of gravity provides a stable spatial reference.

The link between the two spatial references in this system is simple and direct as shown in the classical illustration (Fig. 3.6). The direction of the sun in relation to the hive represents exactly vertically upwards. An angular direction of eighty degrees to the left of the sun is transformed to eighty degrees to the left of vertical on the comb.

A splendid model, and one of Karl von Frisch's most important discoveries, explained in a single concept: The bee dance changes its appearance with the locality of the food site and with the sun's daily passage.

Many After VON FRISCH Work with Exact Angles, Few Reflect on Error Sources

Von Frisch was not alone in reporting exact angles down to the first place after the decimal point. One can find such exact values in many following studies of other authors. Angles down to the first decimal place is either, as in the case of von Frisch, a directly measured value, or the result of a mathematically calculated mean.

Fundamentally, there is a danger of reporting values more accurately than actually recorded, or unreasonable in the context they were taken. The assertion of an angular accuracy higher than measurable or reasonable suggests that the dance offers correspondingly precise directional information and can even be used as such by the recruits. The success of the communication, that is, the arrival of the recruits at the goal, would appear to confirm such an exact description of the site location.

In his publications [95, 96], James Gould also reported highly accurate angular values. He did not comment on the margin of error and so on the reliability and accuracy of his angular measurements taken from video recordings, but identified two forms of systematic error in recording angular values. The first was an experimental error. The lamp, placed in the hive as an orientation reference, was at a determined distance from the comb. In his experiment, bees danced over a comb area of about ten by ten centimetres. From this area, bees viewed the lamp that afforded a reference for the un-manipulated bees instead of gravity (see Figs. 5.5 and 5.6). Because the lamp appeared at a different angle from different positions on the comb (parallax), Gould reckoned an error of 1.14 degrees of arc (here with an accuracy of one-hundredth of a degree) in recording the angle indicated by the dancer.

What is the point of such accurately recorded angles considering all other coincidental errors influencing the measurements? They suggest a measurement performance that is not attainable.

As the second source of error in recording the dance angle, Gould refers to a systematic mistake made by the bees, the so-called misdirection. Misdirection is a deviation of the danced angle from the reference value, that is, the angle perfectly aligned with the goal [83]. External factors are responsible for misdirection.

Gould assigned a value of 1.4 degrees for both error sources together and subtracted these from the dance angles in the video recordings [95, 96]. Here, Gould worked with an accuracy of one-tenth of a degree.

Most Recruits Find the Food Site Despite the Misdirection

In studying the accuracy of directional information in dances, von Frisch and his co-workers compared the recorded angular value with the actual direction of the food source as the reference. Misdirection occurred when bees in the observation hive could see the sky. When prevented, the misdirection was minimised but not completely abolished. Martin Lindauer, von Frisch's most valuable collaborator, discovered systematic misdirection that increased and decreased during the course of the day. This misdirection was noted as "residual-misdirection." It could be shown that regular fluctuation of the earth's magnetic field is responsible [99, 151, 152, 165, 266].

The discovery of the perception of magnetic fields by bees as an unexpected by-product of the communication research on honeybees was another achievement in a long line of important insights contributed by von Frisch and his co-workers.

Observation of flight behaviour of honeybees [64] shows that these insects also react to the solar wind, that is, streams of electrically charged particles that at times of massive eruptions stream out from the sun into space. Strong solar winds disturb the orientation of bees. It remains to be shown how much the dance behaviour is affected by this interference.

Fan experiments were carried out for situations with large misdirection (those measured from the waggle dance deviate up to forty degrees from the direction to the food source) to see if misdirection disorientated the recruits. In no single study did recruits follow the misdirection. Instead, the greater majority flew to the control station in front of the feeder visited by dancers that had given the wrong information in their dance [83]. There are no details given in this study about the number of recruits that flew to the feeder, also visited by foragers, during the observation period.

Von Frisch intimated that ignoring the false information was consistent with the accepted meaning of the dance communication: In his opinion, the recruits corrected the falsely indicated direction because in the hive, they were subjected to the same error influence as the dancer [78].

Such an error compensation must be exact, because with increasing distance a particular angular error leads to an ever-increasing lateral displacement from the goal. A misdirection of ten degrees of arc for a goal five kilometres away, if not corrected, would lead to laterally missing the goal by 800 m; with a forty-degree misdirection, the lateral deviation would be around 4.2 kms. Were the dance followers to obey the instructions, they would never arrive at the site. But they do arrive.

A five-kilometre flight distance is by no means the outer limit for foragers. Flight distances of a bee colony are dependent essentially on the distribution of blossoming plants [12, 229], and up to ten per cent of foragers can cover distances over 9.5 kms from the hive [12]. If one takes the role of communication between the bees in the field into account (here again the repeatedly mentioned blind spot), and the consideration of how recruits manage, despite the misdirection, to find their way to the goal, then the three phases of distant goal orientation provides, at the least, an equivalent explanation of the goal finding success. A precise dance and an exact execution of the instructions are not necessary.

The Dance Delivers no Reliable Directional Instructions

As already mentioned, dance figures of an individual dancer can differ from one another and the paths traced by different dancers for the same food site also vary (Fig. 7.3)—the divergence of directional indications become even greater [224]. If a human observer has a number of dances available for analysis that all promote the same goal, errors cancel one another out in the calculation of direction. The more dances and the more false instructions that are included in the calculation of a random data set, the more the errors balance each other out (too far right, too far left) and the more exact the result of the calculation of the goal becomes (see also Fig. 8.1). These calculations are not trivial and challenging. The statistical treatment of vectors, starting with the calculation of means, requires a particular approach. Circularly distributed data have to be analysed according to certain rules; circular or directional statistics is a special area of mathematics [9, 126].

The residual misdirection that changes with the passage of the day [83] is another matter and depends on the position of the dance group on the vertical comb [224]. These errors are easier to rectify because it is not a statistical problem, instead consistent and directly recordable mistaken directions indicated by the dancer.

Von Frisch had already observed that for food sites close to the hive, the waggle phase varied significantly in sequential dance rounds. Waggle phases following the alternatively right and left turns of the dancer are correspondingly directed more to the right or to the left (Fig. 7.4). For this, he introduced the term "sickle dance."

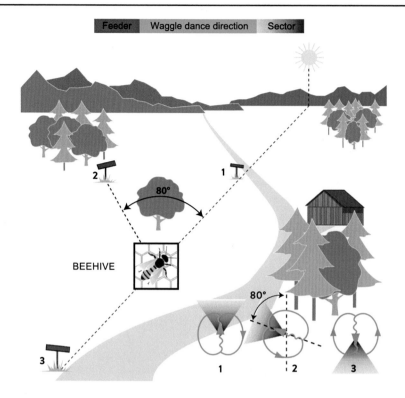

Fig. 7.3 This illustration takes the "half-true" directional instructions from the waggle dance into account. No explicit directions are given to the goal, instead to a diffusely delimited sector in which the goal can be found (see also Fig. 5.7)

The angular divergence of the sickle dance narrows the further the goal and the longer the waggle phase takes. From the distribution of angular values, the statistically calculated direction of the goal becomes ever more exact the further away the food site and the more the data collected [13, 42, 89, 135, 267].

The randomly distributed, inconsistent errors in the highly variable dances and the systematic misdirection set considerable limitations on the usefulness of a site description in the bee dance. Computer models based on the measured data taken from the dances show that to find a goal from the dance information alone is impossible [191, 192].

The unreliable directional information shows orientation signals and cues and communication between bees in the field are unavoidable. Despite the half-truths in the dance communication, they lead the recruits precisely to the goal.

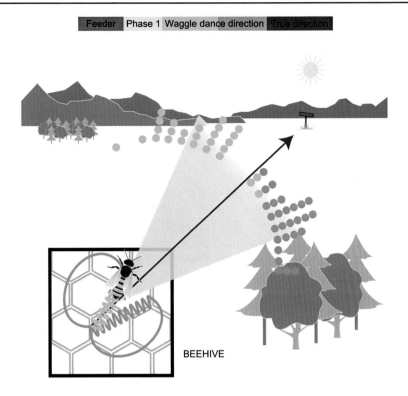

Fig. 7.4 The graphics illustrate the movements and corresponding directional instructions of a dancer promoting a food site 150 m away (after 267). From a right turn, the dancer follows the waggle phase shown in green. Subsequently, she executes a left turn and waggles along the path shown in orange. Another right turn follows and so on. The points in the field give the extent of the variation of the directional information that can be read from the waggle phases. The colour of the dots corresponds to that of the waggle phase paths. The red arrow gives the exact direction of the food site

How Do the Bees "Read" the Distance to the Goal from the Dance?

The orientation of the waggle phase can be read directly from the dance and apparently, in the direction in which the goal lies in relation to the hive. It does not have to be derived from indirect dance phenomena that could contain directional information. It needed the observational skills of von Frisch to recognise this.

The situation for distance indication is different. Should the waggle phase contain encoded distance information, these must be built-in and not decoded as for directional instructions. Results of von Frisch's step experiments led to the conclusion that distance information was included in the dance and to an intensive search for a distance-relevant element in the bee dance.

Von Frisch and his co-workers were the first bee researchers to look for a correlation between elements in the dance and the distance to the food site. The best match they found was the number of dance rounds that occurred in 15 s. They counted how often a dancer turned right (one round) and left while the stopwatch ran. The 15-s observation period was adopted as reasonable after a number of trials, because they found differences in the duration of single dance rounds depending on how far it was to the goal: For close goals, bees dance quicker rounds. Plotting a graph of the results revealed an asymptotic curve over a distance of ten kilometres that became ever flatter. Interpreted this means: Equal stretches of the flight lead to less and less effort in single dances, the further away the destination from the hive. It is therefore not possible to establish a distance-dependent calibration factor that allows every particular flight distance to be calculated. Each additional dance round means an ever-greater additional flight distance. When two quantities are related to one another, one speaks of a correlation. If one of the quantities changes, so does the other. The changes can be in the same direction (positive correlation), or one increases and the other decreases (negative correlation). Correlation may have a causal relationship, for example, deafness and noisy working conditions. Correlations may also have no causal relationship such as the decrease in the population of white storks and the decrease in the birth rate of people (i.e. not what is told to children). In the case of a causal relationship, a correlation has an information value, a rule that allows the quantity of one to derive the quantity of the other.

From these considerations of correlations, research on honeybee communication has a three-stage task. First, it must show that there is a correlation between the flight distance of a forager and one or more of the quantities of the dance figure. Next, it must show that quantities associated with flight operate directly on the parameters of the dance figure. Third, it must show that the parameters dance followers take from the dance translate into flight distance.

For the first step, von Frisch and his co-workers suggested nine elements from their observations of the waggle dances that were conceivable as signals for distance indication and which they sought to correlate with the flight distance to the goal. These were 1. Frequency of the waggle movement; 2. Time taken to complete an entire dance round; 3. Duration of the waggle phase; 4. Time for the return run; 5. Number of waggle movements per run; 6. Length of the waggle phase stretch; 7. Duration of the sound emission; 8. Frequency of the vibration; 9. Temporal construction of the vibration pattern. From slow motion recordings [80], electromagnetic measurements of the movements [51, 55, 248] and sound recordings [282] apparently unsuitable quantities, were excluded from the list. The length of the waggle phase remained as the best-correlated quantity.

The length of the stretch covered over the comb during the waggle stage can also be obtained from the duration of this phase of the dance. Von Frisch wrote the following about the suitability of the length of the waggle phase on the comb:

> The values of the length of the waggle path exhibit such a strong variation that they also can be eliminated [...]. The length of the straight run would also not be an appropriate signal quantity because due to the smooth transition of the curve, the beginning and end of the waggle run is not sharply defined ([78], p. 101; see Fig. 7.1).

Von Frisch was justifiably sceptical, about whether the length of the waggle phase was suitable as a useful criterion for distance. So it was that no waggle dance length measurements were made in studies following on from von Frisch. All used the time information from video recordings. Alternative technical solutions for an exact determination of the waggle phase duration were applied during the waggle phase vibration pulses, or also simultaneously with observed fluctuations of electrical charges the dancers produce during the waggle movements.

Rodrigo De Marco and co-workers, using video recordings in extensive studies of the waggle phase duration, confirmed a high variability, even for single dancers [42]. From their results, this applies to the duration of the waggle phase and also to the number of single waggle movements during a waggle phase. Both parameters are interesting with regard to how the dance follower can gain information. Do they measure the duration of the waggle phase? Do they count the waggle movements?

Dance followers could also use other parameters of the waggle phase as distance criteria. The number of the vibration pulses in the waggle phase, the sum of the duration of the vibration pulses per dance round, the duration of the waggle phase, and the number of the waggle movements are all correlated [156].

Scientifically, the site information readable from the dance figures becomes ever more precise the more extravagant and complex the entire analytical procedure. The goal instructions are more accurate if one omits the first and the last dance rounds of the observed dancer and includes only the dance rounds from the middle [34].

To propose that bees are in a position to undertake a similar complex analysis is a bold hypothesis.

Are the Number of Waggle Movements and the Duration of the Waggle Phase Useful Signals?

The relationship between the duration of the waggle phase and the flight distance is linear up to a distance of several hundred metres; longer flight distances correspond with longer waggle phases. This is different and simpler than what one gets from the number of dance rounds in fifteen seconds. This simple connection though does not hold for the outer range of the foragers. The linear curve describing the waggle duration with increasing distance flattens out and asymptotically approaches the axis representing distance.

The initial linear section of the distance curve would permit, in principle, a universal calibration factor allowing a simple determination of any flight distance either from the duration of the waggle dance or the number of single waggle movements. Resolving the duration of the waggle phase, or the number of waggle movements, is irrelevant because they are equivalent; regardless of the flight distance, the dancer executes about fifteen waggle movements per second [51].

However, considerably different values are given for such calibration factors. Harald Esch [52] determined that for a flight distance of 200 m a single waggle movement implied 33 m, and 77 m for a flight distance of 2000 m. The values given by De Marco and Menzel [40] and De Marco and co-workers [41] (26 m per waggle movement) are close to those of Gould [95] with 24 m per waggle movement. Greggers and co-workers [107] find sixty metres per waggle movement. In many publications, no independent calibration factors are determined and employed, instead the flight distances corresponding to the measured time are taken from von Frisch's classical curve [78].

The assignment of a particular flight distance to a single waggle movement is also temperature-dependent [52]; a universal calibration of the distance information is impossible.

Bees cannot simply "calculate" the flight distance with a rule. Neither counting the waggle movements nor determining the duration of the waggle phase provides one with reliable information about the distance to the food source.

There is another serious problem. The asymptotic nature of the curve representing the relationship between the flight distance and the waggle dance duration leads to a difficulty of which von Frisch was fully aware: dance rounds of different lengths indicate flight distances of the same length. Equal time increases in the dance correspond to ever-increasing stretches of the flight.

To determine a flight distance from the dance at all on this basis attributes remarkable abilities/characteristics to bees. One assumes that different flight distances appear to bees to be the same because sections of the flight more distant from the hive are stored as shorter in their memories and distances at the start of the flight, more heavily weighted [81]. Von Frisch formulated this complicated hypothesis as follows:

> The characteristic SHAPE OF THE DISTANCE CURVE can be understood by assuming that during their flight, of the many overflown landmarks per time interval, bees "forget" a small number [...]. If one takes a "memory loss" constant for the moment of memory that vanishes per landmark at the start of the flight as "ascending", so it is possible to calculate a relationship between the path flown and elapsed time that corresponds with the experimental curve. ([78], p. 127, emphasis in the original).

That the determination of the flight route and distance must be based on a complex memory performance would be taken as likely in many studies (e.g. 27, 47), but no one found a mechanism like that proposed by von Frisch.

Bees Have an Optical Kilometre Counter

Next to the olfactory sense, the visual sense of honeybees is of paramount importance for orientation on their flights. Our knowledge about the optical cues flying bees use to orientate was thoroughly and excellently summarised and published by Miriam Lehrer in an anthology in 1997 [149].

Forager bees get an impression of the distance covered from their visual sense.

Data for the correlation between dances and the flight distance gathered by different researchers differ substantially. For the same distances, the authors report differences of 300 per cent in the measured duration of the waggle dance. For data gathered for the distance the differences result from—in contrast to the directional information—not solely due to the behaviour of the honeybees, instead also from differences in the external conditions of the study. This sounds complicated—what is meant?

Honeybees measure the covered distance with their eyes. The visual structure of the landscape in which studies are conducted influences the result to a certain extent.

Research on the "kilometre counter" of honeybees not only produced an explanation of the difference between the data from different studies but also provided the reason why a universal calibration of the distance indication for honeybees is, in principle, not possible.

Investigation of the bee kilometre counter led to an understanding of the causal relationship between what bees experience during their flights and then express in their dances and how dance followers use this correlation. The studies unravelled the honeybee kilometre counter [57] and produced the first evidence that information recruits observe in the dance, alone influenced the execution of the first flights of recruits towards the goal [59].

Von Frisch and his co-workers proposed the energy requirement for a flight as a measure of the flight distance. Bees that flew against the wind and indicated a longer flight distance supported this. Similarly, bees to which a weight was fastened and those that had to fly up a steep incline to the food source also indicated longer distances. These studies revealed that it is the outward flight to the food source and not the return—and not the combined stretches—during which bees determine the distance to the food source [78].

Today, we know Bees possess an optical kilometre counter. Images of the landscape a bee flies past move across its compound eyes and convey an impression of how far it has flown [57, 59, 243, 262].

The explanation is simple, and a good example of how correct data can lead to completely different interpretations. Bees flying against the wind, with extra weight or uphill, move close to the ground in all three situations. The optical flow experienced by the flying bees increases and with it, the subjective impression of the distance covered. The waggle phases become longer.

The following from my publication in 2014 [247] explains the essential details of the optical odometer of honeybees:

Fig. 7.5 A forager bee that flies through a narrow tunnel with patterned walls to a feeder is deceived by the resulting optical flow giving her the impression that she has flown a much greater distance than she has in fact covered

The origin of the now well-known tunnel experiments with honeybees related to their measurement of distances is introduced here because it leads to an understanding of how chance can play a significant role in scientific breakthroughs. From an earlier research visit to the Australian National University in Canberra as a guest of professors Mandayam Srinivasan and Shaowu Zhang, I was aware of their studies using a tunnel with optically patterned walls to explore how exactly bees could detect sites where they had previously been rewarded with food. I was also familiar with the older publications of the German zoologist, Harald Esch, who with his co-workers presented the first evidence that distance measurement by honeybees was not, as supposed originally, through the consumption of energy during flight, instead through so-called optical flow. He compared the length of the waggle phases of bees trained to fly from the roof of one tall building across to that of another with bees that flew the same distance but closer to the ground. Optical flow decreases the further bees are from the ground and the dances are shorter. The way it is with ideas, one day it occurred to me to ask why the Australian colleagues did not observe the dances of the tunnel bees, if, under the conditions of their experiments, these occurred because with free flying bees this provided an indication of the distance they had covered. The dances should after all, reflect a subjective impression of the bees. In an initial telephone call with Canberra the question came: Can that be done? How? What is needed? In the previous years, I had learned from Martin Lindauer how to employ observation hives and I suggested this to the Canberra colleagues. To bring them rapidly forward in this new area, I asked Heinrich Demmler, our beekeeper and cabinetmaker at that time, to build an observation hive after the original model used by von Frisch. This was then carefully packed and sent from Würzburg to Canberra. Monika Altwein was engaged for the instruction of the Australian colleagues in the employment of the observation hive and measurement of dance parameters in the tradition of Karl von Frisch and Martin Lindauer. In the Spring of 1999, Monika Altwein followed the observation hive to Canberra. The German-Australian collaboration resulted in the unequivocal confirmation of Harald Esch's ideas and for the first time enabled quantitative data to be obtained on the function of the bee optical odometer. These studies calibrated the odometer. They confirmed to what extent distance indicated in the dances was dependent on the optical structure of the environment. Experimentally derived distance graphs for the analysis of dances are therefore valid only for the specific areas in which they are measured. These results were published in SCIENCE. For cream on the cake, I conceived a cover illustration for the edition that would represent the basic concept of the experiment. I delegated this task to a co-worker of the BEE group in Würzburg who produced an illustration that won first prize in the competition "Visual Science" in the year 2000 (Fig. 7.5). Thinking about the successful experiments it occurred to me that it should be possible to employ the tunnel bees as "lying" bees. Such "lying" (i.e. misled) bees would indicate a distant goal although in the tunnel they had only flown only a short distance. In other words, one could now test if recruits that followed the dances of tunnel bees, searched for food at a distance corresponding to a distance the length of the tunnel where the dancers had found food, or would be caught at the greater distance indicated in the dance. To acknowledge Harald Esch's unprecedented contribution I suggested to my partners, Srinivasan and Zhang that we invite him to join us as an honorary guest in Würzburg for the experiment. In the resulting studies, we could indeed show that the "lying" (misled) bees were taken seriously. Searching foragers and their followers were caught in nets in the area that dancers had advertised. No bees were caught in front of, or beyond this area ([59], p. 62f).

Waggle Durations Are Dependent on the Landscape Through Which Bees Fly to the Food Source

Because the visual odometer depends on optical flow and not on distance, the impression a bee has of the distance covered is determined by the structure of the surroundings. A bee flying over an evenly mown field or over a diverse and contrast-rich landscape indicates stretches of similar actual length with dances of different lengths. A highly structured landscape stimulates the bee odometer more strongly than a uniform one; the duration of the waggle phase increases with the higher optical flow. However, bees flying to food sites the same distance away, but which lie in different directions, could result in waggle phase duration, and so distance indications that differ by a factor of two or more. A waggle phase of 500 m could mean a distance of 250 m for a flight to the south and 500 m for a flight from the same hive west [59].

Bees that fly through a patterned tunnel and are trained to feeders at different distances from the hive exhibit differences in their waggle phase times that cannot be explained by optical flow alone. Is there after all an additional distance-measuring component? There are still many open questions.

However, the dependence of the waggle phase duration on the optical features of the flight path from which the dancer derived the distance and which she employs in her dance leads to the need for an interim hypothesis related to the recruitment behaviour of honeybees: Recruited foragers must take exactly the same flight path and altitude as the dancer through the landscape. Only in this way can they experience the same optical flow as the dancer and so correctly apply their perception of the distance information.

If one considers the recruitment to a feeder as a three-phase distant goal orientation, the interim hypothesis is unnecessary and inexact dances are adequate. The dance is deployed only in the first phase.

Bees Speak Different Dialects

Karl von Frisch included studies in his publications on distance indication in different races and species of honeybees [78]. The temporal structure of waggle phases of all investigated bee races and species alter in length with increasing distance to the food site, but at different rates. The same times in the dances indicate different distances. Bee races and species were proposed to express dialectical colouring in their dance language [19].

If such differences result from different optical flow during flight to the food source, because the landscape structure exhibited corresponding differences [238], is unclear, or if dialects in von Frisch's understanding in fact exist, or his observations merely indicate different experimental conditions. A similar more recent study in which the foragers of three different bee species were underway along the same flight path and in the same locality confirmed the existence of dialects and were discussed as an adaptation to the prevailing environment [141].

The transfer of visual impressions during the flight to a food source to the duration of the waggle phase is determined by the structure of the landscape and is different depending on the race or species of the bees.

Bees Are Aware of the Passage of Time

Von Frisch not only noticed that the dance figure changed with the location of food sites but also that for stationary food sites dances also changed with time. This has to be so because the sun, as the reference point for orientation in the field, moves in relation to the hive. The bee's sense of time ensures that the sun's movement across the sky and the consequent change in the direction are indicated in the dance figure, even when significant time elapses between the foraging flights and the dance behaviour [202, 203].

Natural scientists recognised early that bees have a sense of time. The basis for this was the observation that flowering plants open and close their blossoms at different times of the day. The renowned Swedish natural science researcher Carl von Linné (1707–1778) designed a flower clock: For each hour of the day, he found appropriate species which were then arranged accordingly in a round flower bed— the time could be told thanks to the frequency of the visiting bees [154].

As an adaptation to the temporal differences of the flower open times, bees have extended their ability to couple the sense of time with their knowledge of the area and to learn when and where there is something for them to harvest [195, 96, 290].

Bees and Vectors

8

A reductionist mathematical terminology and way of thinking can also lead to applying abstractions, hypotheses, and models in biology. In the case of the dance language, this approach emphasises the importance of dance information.

The concept of vectors is central to analytical geometry. Arrows indicating a direction and their length denoting a numerical quantity represent the values of vectors. Extending from an origin, the arrows are drawn in either a plane or space. Vectors accomplish, in the form of polar co-ordinates, what Cartesian co-ordinates achieve in establishing points in a plane or space on their corresponding axes. A vector that indicates a particular site is a position vector. The application of vectors would appear to be perfectly suited to research on dance communication in honeybees, where direction and distance to a goal bees visit are concerned.

Foragers that visit a feeder supply the data for an appropriate vector calculation for bees. Individually marked bees returning to the hive were recognised and their dances measured (see Chap. 7 in relation to the question about accuracy). A vector was calculated for the movement pattern of each dance (direction and duration of the waggle phase) marking a supposed goal in the field. The waggle phase indicated the direction and the distance most often taken from von Frisch's classic curve ([78], Fig. 61) or calculated from a calibration factor.

The formal basis of calculating vectors from bee dances and identifying the goal is enabled by modern data gathering and processing methods. These open the possibility of automating the collection of the relevant dance parameters, calculating the goals from these, and plotting them on maps (see Fig. 8.1, [200, 276]).

Establishing vectors from observations and measurements of the dances, goals (points in the landscape) derived from the values can be plotted. Initially, these exist as sites only in the minds of the researchers, but allow predictions from basic assumptions about bee communication to be tested.

What are these predictions? The central and most important prediction derived from the dance language states the following: Recruits arrive at the goal promoted in the dance. Is this the case?

Recruits arrive at locations indicated by vectors calculated from the dances, only if experienced bees that had danced in the hive and promoted the sites visited these, or fragrant flowering plants were present there. Otherwise, in such cases, the goal orientation signals and cues are absent.

The classical explanation of an absence of recruits at such goals is the argument that they do indeed follow the instructions in the dance and arrive at the site but are not motivated to land and so not observed. This notion is also included in the classical view of the bee dance (see Fig. 3.6).

If one calculates a vector from a dance angle and an indication of distance, one obtains a goal point. In fact, however, errors in recording and the variability of the dance even in the best case permit divergences from the true goal to be obtained. De Marco and co-workers conducted detailed analyses for the variability of dances indicating the same goal. From a complex calculation, it was found that for a goal 215 m from the hive, dances with at least forty waggle movements still diverged by at least five degrees. This corresponds to a lateral error of eighteen metres at the goal. The angular error increases if fewer waggle movements are included in the waggle phase and if the recruits follow fewer dance rounds [42].

The Bee Dance Describes a Goal Area, not a Single Goal Point

Instead of a single goal, each dance must indicate an area within which, with a certain possibility, the actual goal lies. The strict mathematical vector treatment brings a result with a clear and formal accuracy that does not correspond with the unclear raw data from the dance.

Translating the direction and distance of each single dance figure measured with mathematical precision into vectors and then determining a fictive goal point in the field from the vectors reveal that for the unchanged position of both hive and feeder, these calculated goals are widely distributed over the landscape (see Fig. 8.1).

The situation in the exemplary illustration in Fig. 8.1 can be considered from three separate viewpoints: 1. One cannot transfer these fictive goal points to the landscape, but instead takes them as "information noise" the recruits must cope with to reach the advertised goal. 2. One accepts the points for a real location in the field that can be derived by calculating the mean and is targeted by the recruits. 3. One assumes that the dance information does not indicate the location of a goal point, but instead leads to a goal area in which there is a high possibility to find goal orientation signals and cues—one sees recruitment then as a distant goal orientation.

The first two viewpoints provide a good example of the completely different opinions about the dance language of honeybees, in which two fundamental viewpoints in biology turn up again. Only in the life sciences is it sensible to ask the

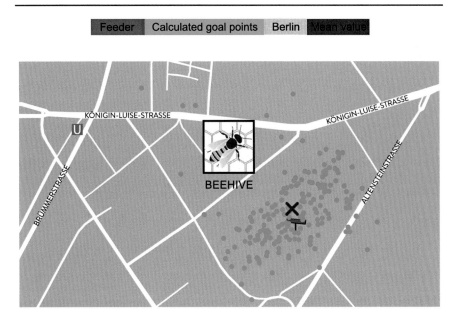

Feeder   Calculated goal points   Berlin   Mean value

Fig. 8.1 The illustration shows a beehive in Berlin. Bees were trained to a feeder (marked orange), 234 m away from the hive. The goal points read from 192 dances are marked in blue (after 200). The red cross marks the point representing the statistical mean calculated from all the points. The wide distribution and inaccuracy of the dance information are clear

question "why" in two different ways and to provide an answer. The unclear "why" can on the one hand be asked as "how." In this case, one is interested in the causal connections, the mechanisms that are the basis of the phenomenon. On the other hand, one can understand the "why" as "for what." Here, the interest focuses on the usefulness of the events. This makes sense only in the life sciences; one would not ask a physicist about the usefulness of the laws governing a pendulum. Nevertheless, the answers to both questions about the same phenomenon should not lead to a contradiction if both are correct.

Researchers Calculate the Mean.

Do the Bees?

Two facts, the broad distribution of calculated vectors and the exact arrival of recruits at the feeder for which the dancers advertise, constitute a problem for the classical view of the bee dance (Fig. 3.6). The question here is about the mechanism of events that ensure recruits arrive at the goal, in spite of inaccurate instructions. Researchers faced with the problem do not transfer the fictive end points one to one into the field, but instead ask how recruits despite the "noisy" information land exactly at a goal a dancer has promoted. In the end, a single vector indicating the correct direction and distance should bring the recruits to the goal. What other solutions have the bees obviously found?

The old master and researcher of the bee dance, Karl von Frisch, had already noticed the above-mentioned problem from his direct observations of the dance:

> Bees sent out from the dance fly to the goal with greater accuracy than would be expected from the spread of the individual dances. [80, p. 263]

How is it possible that recruits land precisely at the site from which the dancer came and advertised, when the dances actually contained "sloppy" information? His solution:

> One can conclude that while following the dance they take an average of the many single values. [80, p. 263]

Many bee researchers accepted this idea and in fact, the averaged value of many dances indicates the correct goal, the more accurately the more values that are included (see Fig. 8.1).

Such a result, the calculation of the correct target using statistical methods, is to be expected when the single values are symmetrically distributed around the (correct)) mean value. A marksman who misses a target by the same amount an equal number of times to the left and to the right could, statistically, be said to have hit the target exactly in the centre. Nevertheless, better not to check the target. If one accepts that recruits average the information in the dance to obtain the correct result and so reach the promoted goal, one must introduce another interim hypothesis. The single value recruits obtained from the total provided are distributed equally as errors to the left or to the right; otherwise, the calculation of the means will not give the correct direction.

Even if the calculation of a mean from many single values provided reliable information about the goal, and the more accurate the greater the data set, such a statistical view is irrelevant for a forager that sets out to find the goal promoted by the dancer, she is aware of only a small, random selection of all the dances. The recruited bees have no idea of how correct or incorrect the data are. For them, the dances are "half-true."

Is It to the colony's Advantage if Recruits Miss the Goal?

Some researchers question the biological significance, the "for what," of the vector spread and transfer the calculated goal points as real goals correspondingly distributed in a landscape that is accordingly visited by recruits. They do not regard the broad spread of goal points as a problem, but instead as an adaptive gain for the bee colony. It follows that they also do not propose a hypothesis to explain how recruits in spite of unclear instructions arrive at a particular site by calculating it from scattered information. Their hypothesis focuses more on how not reaching a narrowly defined goal could advantage the colony.

In their view and research approach, the spread of goal points across the landscape does not assume that information contained in the dance leads recruits to a single goal; instead, they postulate recruits accept and follow all the different goal indications. This should assure that recruits distribute themselves correspondingly through the area. In this concept, the high variability in the dance is an advantage for the entire colony [89, 265, 267, 281]. If the distribution of bees is spread optimally across a wide area, this would ensure that foragers of a colony would not harvest too close together. The colony as a whole would profit because many new flowering sites would be visited. The inaccuracy of the dances fulfils this aim.

Precisely following divergent instructions from the dance served as the basis for computer modelling [191, 192] that calculated the advantage a bee colony would have to a corresponding distribution in the field. Observations of dances came directly from the hive and events in the field derived from computer model calculations; no actual field observations or measurements were available for a comparison (while, as already said, this area is the blind spot of bee research).

Ever fewer in the profession are of the above opinion. Instead of accepting that the inaccuracy of the dance fulfils a role, many bee researchers today believe that the dancers are simply unable to dance precisely [199, 250, 251, 253]. The fact that dances vary more or less according to which stimulus outside the hive the dancer uses as the reference also supports this opinion—the variability of dances in reference to a light source is less than that in the dark hive where the dances are referenced to gravity [251, 253].

The growing doubt over the hypothesis that inexact dance communication is useful makes another associated hypothesis unnecessary. Should both proposals related to the bee's response to the noisy instructions about the location of the goal agree, then bees must either determine the mean of the dances or fly to a particular goal that the dancers also visit, or they decide not to calculate the mean and distribute themselves over the area.

It is also possible that neither of the two explanations is correct. Assuming that the rough locational information from the dance is correspondingly vaguely helpful for the recruits, they could find a specific goal in the field only through coupled stimuli. The scientific dispute about the significance and meaning of inaccurate dances would no longer have a foundation if one considered the "half-true" information from the dance as the first phase in a chain of distant goal orientation events. Subsequent more accurate information in the field leads the recruits to the advertised point. The recruits would then not have to calculate means nor optimally distribute themselves over the area. It would be sufficient for them to fly off to the roughly indicated goal area and then use whatever help they found there to lead them to the goal (Fig. 8.2).

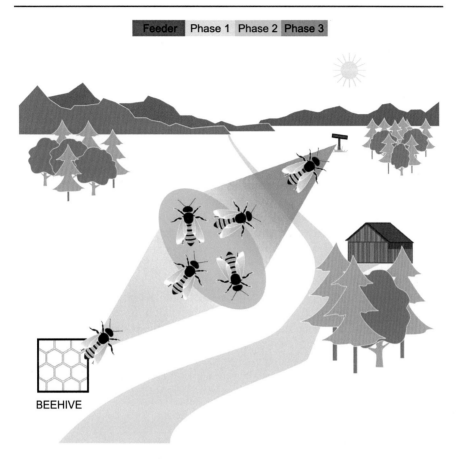

Fig. 8.2 Bees that fly to a food site new to them, that another bee had advertised in a dance, pass through three phases of distance orientation to reach the goal. 1. They set out in a direction, following the information in the dance (SEND—yellow sector) and orient themselves in relation to the sun. 2. They arrive within a search area (SEARCH—green sector), the spread of which is determined by external and internal factors of the recruits; they come across stimuli that lead them to the goal, the scent of flowers and the signals from the dancers that continue the communication in the field (ATTRACT—red sector)

Bee Dance Data and Ecological Research

Localities read from the dances of bees are fundamental for studies focused on the pollination of flowering plants by honeybees, in particular, crops in agricultural lands. Reading the dance figures can reveal which areas the currently studied bee colony visits and how the geographical pattern spatially and temporally changes during the year [15, 34, 35, 36, 224, 249]. Pioneering studies in this area arose again from von Frisch's research group, authored by doctoral candidate Herta

Knaffl [140], who was the first to employ the concepts about the dance to gain an impression of the foraging areas of a bee colony.

Here too, a supposed projection of the goals read from the dances takes place in the landscape. Von Frisch's classical model provides the basis for this (see Fig. 3.6). The statistical treatment of measured values taken from the bee dance provides averages of the direction and distances that lie very close to the goal point (see Fig. 8.1). These means are the more reliable, the greater the data set included in the calculation.

For a consideration of half-truths in bee dances, the Bayesian statistics are better suited than conventional statistics that only question if an hypothesis is true or false. In contrast, the Bayesian statistics examines the probability of an infinite number of stages ranging from "false, unlikely" to "true, likely" and so permits half-true statements. A certain form of simulation (Markov chain-Monte Carlo simulation), combined with a Bayesian view of duration and direction of the waggle phase, results in a particular probability distribution for the position of the goals for which the dancers advertise [226].

Rules for calculation (algorithms) developed on this basis permit the most accurate possible determination of site locality. Omitting the first and last rounds of the analysis of single dancers gave the best results for a known goal [34].

The use of site location data obtained in this way is useful and valid for ecological studies. However, it should not lead to the assumption that dance followers are in a position to arrive at such reliable results.

In corresponding ecological studies, the calculated goal points fuse into a goal area [188, 189], again not a specific site location. Through exaggerated calculation methods, the circle closes and one accepts an area, like that contained in the inaccuracy of waggle dances.

RoboBees and Radar Techniques

9

The idea of vectors in communication research of honeybees and the associated formalisation and mathematical treatment of notions and research of recruitment behaviour of bees led to a series of interdisciplinary questions, approaches, and research projects.

Comprehensive thinking often brings new ideas and hypotheses. In this case, a new methodological approach in an attempt to verify the dance language hypothesis and communication of the goal direction (Fig. 3.6).

RoboBee Has the Honour

It is easy to understand why the pioneers of formal communication research had the idea, before biologists, about information content in waggle dances of honeybees [see p. 168ff.; 113], and to apply technology to explore information transmission from a dancer to a dance follower. Their notion: Employment of an artificial mechanical dancer would allow controlled amplification of transferred information and an investigation of whether dance followers are, in principle, in a position to receive more information than the "half-true" dance movements real bees can send. The transmitted information could be increased and uncertainty in highly variable natural dances replaced by noise-free, clearly defined "standard dances". In addition, it would allow an investigation of which dance parameters are relevant in the communication.

The conceptual beginning for experiments with artificial mechanical bees, in all published investigations, was the dance language and its assigned key role in discovering a goal in the field. The aim of these studies was to send recruits to points in the field for which identical, reproducible dances of mechanical bees provided the direction and distance from the hive.

© The Author(s), under exclusive license to Springer Nature Switzerland AG 2022   95
J. Tautz, *Communication Between Honeybees*,
https://doi.org/10.1007/978-3-030-99484-6_9

Movements carried out by mechanical dancers are precisely controlled, such a RoboBee dance does not vary and contains no uncertainty or lack of clarity in the message it sends. The waggle phase, in contrast to that of a living honeybee, can indicate the direction to the goal within one degree of arc.

With a functioning mechanical dancer, it must be possible to fundamentally test if dance information is sufficient to bring recruits to a goal, as the central hypothesis proposes [38, 49, 50].

A methodological requirement for success in the employment of artificial dancers is to imitate signal paths, relevant to the transfer of information between dancer and dance follower.

The research must begin by getting a mechanical bee to release the same behaviour in hive workers as that induced by living dancers. The construction of increasingly complex artificial bees during the efforts of scientists over decades reflects the practical problems against achieving this aim.

A model possessing all relevant characters of a living dancer should function and send recruits to a goal (Fig. 3.6—the opinion that a dance language is implicated), or within a goal area (Fig. 8.2—the opinion that distant goal orientation is implicated).

First Attempts with Artificial Dancers

Wolfgang Steche took the first step with a simple object the size of a bee that he moved around among bees on the comb [246]. Karl von Frisch's comment on this experiment:

> He thought that he had managed to produce directed flights, however his assertions were very inexact and later repetitions were without result. [78, p. 103].

It is remarkable that this severe judgement, or in a milder form also applied to all similarly constructed experiments in the following 50 years, as we will show here.

Esch [53, 55] took a significant step forward. For his models (bee-size pieces of wood or wax-coated dead bees), he constructed a system to move them in an imitation of the dance. The apparatus consisted of a motor, a mechanical template to guide the model through a dance figure over the comb, and an electromagnetic drive to imitate a waggle movement of 15 Hz. The most reliable reaction that Esch evoked from foragers was an attack on the model—bees attempted to sting it [55]. Esch summarised the results of these experiments as follows:

> [The model, J.T.] did not succeed in encouraging hive bees to fly out of the hive to a particular food site. She lacks a critically important characteristic. [55, p. 544].

Esch could not clarify which characteristic was missing.

Gould [95] in the framework of his dissertation employed an artificial dancer that moved appropriately and was somewhat more complicated in comparison to its forerunners. A thin tube projecting from the anterior of the model could offer food samples to workers and so imitated an important behaviour of a dancer. In his efforts to imitate the dancers as nearly as possible, Gould included two further stimuli in his model experiments: Vibration of the comb and applying odour to the models. These experiments were also unsuccessful in recruiting bees to a food site.

Axel Michelsen and his co-workers, several years after Gould, took up the challenge to develop a functional mechanical dancer. Their model, a wax-coated brass body, possessed as an added development wings cut from razor blades. Including all that Gould's model offered, near-field sound with a frequency of 28 Hz and air currents were also produced. The intensity and temporal pattern of these corresponded with that measured by a living dancer. This model was employed in an observation hive over many summers, while at control stations in the field the number of bees seen in the vicinity was recorded [74, 175]. Multiple sightings of the same bee could not be excluded, which, given the low numbers involved, led to statistical problems with the results and significantly weakened the impact of the conclusions. More serious was the clear absence of dance followers in the hive. In the publications and accompanying video recordings, the observed and described reactions of workers to the model were limited to attacks [172, 174, 175]. A model can only take part in the dance communication when it succeeds in releasing a normal dance following behaviour.

"RoboBee" the First Artificial Bee to Succeed with a Bee Ballet

An interdisciplinary research group in Berlin came considerably further with efforts to develop a functional model of a bee dancer (Fig. 9.1). The ultimate aim of the project, stretching over years, was to track and record with radar, the flight paths of foragers that had followed dances before their flight of a model named RoboBee, in the hive. The study deserves a closer look because this experimental approach could reveal what the dance alone achieved.

The experimental approach using an artificial bee dancer and then radar tracking a (real) bee on her outward flight belongs to the most complex and impressive studies so far on communication in honeybees.

In the initial stages of the project, researchers, like those before them, developed a model that could trace out a path over the comb imitating the idealised movements of a dancer. Dance parameters of living dancers, recorded with automated motion detection systems, were analysed, reduced to mathematical representations and employed as movement patterns for RoboBee [145, 275, 276]. RoboBee could imitate much that characterised a living dancer. It provided small samples of food, produced vibrations in the surrounding air with wing-like surfaces, and was warm. In addition, RoboBee released a relevant signal—the odour of peppermint, identical to that at the artificial feeder—and that induced workers to visit feeders familiar to them, to which they flew when they had not followed RoboBee in their dance. This was a further confirmation of Karl von Frisch's discovery of the importance of

Fig. 9.1 Within the framework of the exhibition "From the idea to Understanding", the DFG presented outstanding research projects including the RoboBee. In this project, using an artificial dancer in order to study the behaviour of the dance followers, the flight paths of four bees were recorded after they had followed a RoboBee dance [144, see p. 155 ff]

flower scent: Flower scent on a dancer, or simply introduced into the hive, induces experienced bees to fly out to previously visited sites with the same scent.

Investigations now began to see if RoboBee could induce living bees to use information from its dance and visit the advertised goal.

With a technical variation of the experimental setup, the Berlin group succeeded where others had failed with models: RoboBee attracted forager bees in the hive that moved with it [144, 146]. In all, 78 ballet figures were observed with both RoboBee and living bees in which a single bee followed RoboBee's movements. Because living dancers attract up to ten followers, the challenge to develop a better model still exists.

Do Bees Follow the Instructions of RoboBee on Their Flights?

The research group then posed the question of whether recruits followed the information from RoboBee and flew to sites indicated by the mechanical bee. Three goals for recruits were set up in the field: Two feeders with peppermint-scented sugar water and a virtual goal that did not exist in the field but was advertised by

RoboBee's programmed dance figure. About a hundred foragers were trained to the two real sites [144, 146].

To observe and distinguish between different bees, initially, 193 workers (trained to the site, including bees recruited by them) were marked individually. Six of the marked groups were later observed to be RoboBee dance followers. The flight paths of four of these dance followers were recorded with radar on their flight after the dance behaviour [144]. The results from these four bees are as follows:

Bee Ng 71 followed RoboBee's dance programmed for the virtual, but non-existent feeder. Her flight out of the hive after the following behaviour was not to the virtual goal but directly to a food source well known to her. Bees Ny 18 and Ny 47 also flew, after following RoboBee's dance directly to previously known food sources.

Bee Ng 62 acted differently. Her behaviour was the most interesting from the conceptual viewpoint of the study. She was the only bee to fly directly over a site not previously known to her in the first flight after following RoboBee's dance. Her complete excursion from the hive proceeded as follows: Ng 62 flew over all three sites, one after the other and without pausing: at first over the feeder in the field that RoboBee had indicated and which she had not previously visited, thereafter the virtual, but non-existent site for which RoboBee sometimes danced (but not on this occasion) and finally to the feeder to which she had been trained [144, 146].

The published accounts of all these experiments with radar tracking provide no information about the activity of the around one hundred foragers that were trained to the feeders, nor about foragers they had recruited (their number is not known) or all other bees, in total several hundred, that could have been underway outside during the study period.

The overall situation in the field was complex because bees familiar with the feeder, without having danced with RoboBee and induced to fly out of the hive [144], could also have been present in the vicinity. The possibilities for communication between the radar tracked bees (including Ng 62 with the particularly interesting behaviour) and all other bees on control flights between hive and feeders (dismantled during radar measurements) should not be ignored in the interpretation of the four recorded flight paths. Foragers familiar with feeders continue to visit them repeatedly, even when dismantled, to check if there is again something there for them. It cannot be excluded that the one bee, behaving somewhat as expected (Ng 62), tagged onto an experienced follower induced by RoboBee to fly to the feeder also visited by Ng 62. Simultaneous recording of flights of other bees could clarify the situation (see also p. 100).

That RoboBee transmitted signals to recruits inducing them to follow a dance and then to undertake flights out of the hive was a significant advance on all the previous models. From the presented data, it cannot be concluded that during the experiment a bee, after following a RoboBee dance, flew directly to a virtual feeder. Improved artificial bees in the future will certainly attract more followers and induce more recruits to fly out of the hive. Perhaps, it would then also be possible, analogous to living dancers (Fig. 9.3), to determine the spread of the sector formed by follower's flight paths. It would then be very interesting to see if the spread of this sector is independent of the variability of the dance figure set by programming the model, or if precise dances of the model would lead to a more accurately goal-oriented flight of the dance followers.

Mathematically exact dances are irrelevant for foragers. They manage with the inaccurate information real dancers offer them.

Radar Techniques Could Revolutionise Bee Research

Radar tracking flying honeybees could play a decisive role in finding answers to many open questions in research on bee recruitment. This approach is capable of closing the blind spot—should it succeed in simultaneously tracking the flight paths of many foragers. If one could achieve this for foragers flying back and forth from the hive and dance, and at the same time the flight paths of the dance followers on their way out from the hive, many question marks would disappear. One could identify the search area given in the dance where recruits come across signals leading them to the goal. It would allow the sequence of distant goal orientation in honeybees to be derived directly from their flight paths.

A beginning has already been made. Previous, promising and highly complex experiments, successful observations, and measurements show for the first time in multiple radar flight path data that, in fact, experienced and inexperienced bees fly together to an advertised goal (Rodrigo De Marco, unpublished, personal communication).

A glimpse of important insights obtained from direct tracking of flying bees is seen in the flight paths of recruits in Fig. 9.2 (from 171). These bees followed a dance advertising a site not familiar to them. Their outward flights begin in a sector within which they all fly in the same general direction, apparently following the instructions in the half-truth of the dance. This also applies to distance, because after a certain stretch from the hive they stop flying in the same general direction. Irregular flight paths follow that could imply a search phase, before the bees take a direct path to the goal. This follows the shortest route between the hive and the feeder, for which experienced bees danced in the hive and continued to fly back and forth. The suspicion: recruits met with experienced bees that brought them to the goal, extending the guidance from the dance into the field.

Such speculation about flight paths is no substitute for an appropriate analysis of the events. However, such data provide the hope that eventually the weighting of various goal orientation signals (flower scent and communication between bees in the field), under diverse environmental conditions, can be studied with these techniques. The dynamics in the assembly and situation-dependent changes in the phases of distant goal orientation, which the schematically and graphically super-imposed flight paths in Fig. 9.2 illustrate, could be examined in detail.

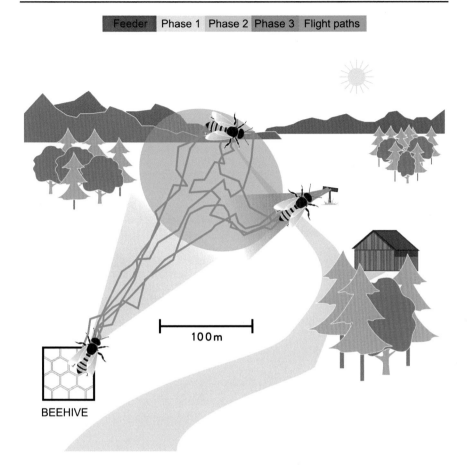

Feeder | Phase 1 | Phase 2 | Phase 3 | Flight paths

100 m

BEEHIVE

Fig. 9.2 Flight paths of recruits on their first flight to a 300 m distant feeder, after they had followed a dance advertising this site [171]. On the last hundred metres to the goal, the recruits were "on track": the shortest distance between the hive and the goal. The three phases of animal distant goal orientation are schematically superimposed on the flight paths (see Fig. 1.1)

Bees Arrive at the Food Source in Groups

Communication and cooperation of bees in the field must begin in the search period, at the latest, to bring the recruits to the goal.

Von Frisch speculated about this possibility but found no way to explore it experimentally. He wrote the following:

> In such an occasion, the question must also be whether when a bee that has foraged from a rich food source, everts the scent organ only when she arrives at the site and swarms around it, or if she keeps the organ everted over the entire way from the hive to the food site and so

perhaps lays a SCENT TRAIL which, although transient in moving air, should, at least on STILL DAYS AND WITH BUSY TRAFFIC, ACT AS A SIGNPOST FOR THE SEARCHING ANIMALS AND COULD BE OF SIGNIFICANT VALUE. Because, in this way, it is possible that newcomers, despite the negative results related to "accompanied" flights, on the contrary, under favourable conditions may be guided by the foragers directly from the hive to the goal. I have attempted in various ways to answer this question. However, we seem to stand here at the limit of experimental possibility and I would prefer to terminate this study with an open question than with an uncertain answer. ([72, p. 172f], emphasis in the original).

Clear evidence of the help recruits receive from experienced bees is that arrival times at a goal are not random, instead a definite grouping of arriving bees is observed. The arrivals of experienced and inexperienced bees follow one another closely. Karl von Frisch described this:

With highly concentrated sugar water feeding one can often see how immediately after a marked bee (one familiar with the feeder), a newcomer lands at the dish. [72, p. 101].

Groups of experienced and inexperienced arrive together at unscented feeders [259]. The recruits circle the feeder after their arrival for up to ten seconds before landing, often on top of experienced bees sitting at the feeder. Arrivals of workers trained to the site become increasingly synchronised over a period of maximally ten minutes. Thereafter, they arrive at the goal only in a tight group. Experienced bees form groups of up to five individuals, joined by up to five recruits. After the second or third group flights of experienced bees, newcomers arrive with them at the site [259].

Gould [95] reported a similar close temporal arrival of recruits at a scented site. He determined the time between consecutive arrival of recruits at a control station (see Figs. 5.5 and 5.6) and grouped the results in time intervals of 0.1 s. 68 pairs arrived at the site with a time difference of less than 0.1 s, 65 pairs within 0.2 s and 35 pairs within 0.3 s of one another. 8 pairs arrived within 1.0 s of one another and 5 pairs within 2.0 s. There is no data for arrival intervals longer than this.

A research group around Fernandez [63], following the pioneering work of Nunez [185], investigated the effect of the frequency portions were offered at a feeder, and eversion of the Nasanov gland, had on the recruitment through dances. They also measured the arrival times of recruits. Again, in this study most of the recruits arrived at the goal within a few seconds after the experienced bees, if the experienced bees had previously danced in the hive and then visibly everted their Nasanov glands at the feeder.

It is not known in any of these studies just where the group assembled on the way from the hive to the feeder. All studies confirm that experienced bees and recruits do not leave the hive together.

However, it is established that recruits arrive at a feeder advertised by a dancer only if this is also visited by experienced bees, even if through certain modifications in the experimental design, in which many goals are advertised, visited, and not visited [280; see p. 211ff.].

The First Flight of Dance Followers is of Relatively Long Duration

Over the longest period in previous bee research, foragers could be observed only at three locations: On combs inside observation hives, at the exit of such hives, often lengthened with a short tunnel in order to see the single bees that walk through it, and at artificially established feeders and control stations (see step- and fan studies).

Typically, for these experiments, a group of individually marked foragers are trained to a feeder (see Appendix for training methods). Researchers have no control over assembling dance followers. This occurs between communicating bees in the dance on the crowded comb. Recruits come from the group of dance followers that set out towards the site advertised by the dancer they followed. Differentiating between dance followers and recruits is necessary because not all dance followers fly out of the hive and not all that do, arrive at the foreseen goal. In some circumstances, a considerable proportion of dance followers, encouraged by the dance, set out towards a goal already known to them. Von Frisch discovered that the flower scent on a dancer attracted bees towards a site familiar to them (but not to the site promoted by the dancer; see p. 23). Usually, they then, after a long foraging time (Fig. 2.5; here more than thirty minutes after leaving the hive), return to the hive loaded with nectar or pollen. Or dance followers fly out of the hive and do not arrive at the indicated goal. These bees, after a fruitless search for a goal at a distance of 200 m from the hive, return after maximally six [56] or more than seven minutes [166] without food to follow further dances in the hive.

Observation of outward flights of bees from a hive lead to an important conclusion: Dancers and dance followers do not leave the hive together on their flights to a food site. This reality is responsible, in the history of ideas in bee research about communication between honeybees, for the convincing alternative of how recruits find the goal, namely the suggestion that "they are sent, not led."

Flight times between the hive and goal are considerably different for experienced bees and the initial flights of recruits. Experienced bees need between twenty and sixty seconds to cover 200 m whereas for this distance, average recruit flight times of 3.1 min were measured [56]. Dietrich Mautz arrived at a very similar conclusion [166]. He determined the flight times of sixty per cent of the dance followers as recruits found the advertised goal. An average of 3.8 min was the result for a feeder one hundred metres from the hive and 3.2 min average flight time for 200 m. Experienced bees cover the one hundred metre distance much more rapidly in 15–20 s [166]. Towne [265] determined for recruits, 69% in his study of the dance followers, an average of about four minutes for a flight path of 200 m, between the first outward flight and arrival at the goal. Gould and co-workers, for goals between one hundred and 400 m measured about six minutes, but from the dance following behaviour, not the exit from the hive. In his investigation, Thomas Seeley determined that recruits needed an average time span of seven minutes from the first flight to a goal advertised in a dance they had followed and to return to the hive, if the distance to the site was about 1,300 m [227]. Von Frisch measured the longest flight time of four hours for a distance of 1000 m between the hive and the goal [72].

All previous studies find clearly longer flight times for recruits in comparison to experienced bees. These conclusions do not suggest that recruits know the location of the goal, derived from the observed dance, at the start of their first outward flight and can fly directly to it.

Ken-ichi Harano and his co-workers chose a very different approach [114]. The research group determined how much honey dance followers in the hive took up in their crops before they undertook their first flight to the advertised goal. They then compared these values with the amount of honey the dancer herself took up in her honey stomach before her flight to the feeder. Above all, the study revealed the following difference between the compared groups: Dancers take up a honey supply directly correlated with the duration of the waggle phase, in other words, the distance to the feeder. Dance followers started with at least double the amount of honey as a supply for the journey to the advertised goal. Other than for the dancer who knows the goal, the supply for recruits does not match the shortest distance to the goal, instead it is adequate for a much greater distance than the minimum required for the shortest flight path. This overloading with fuel is also evidence for the only vague perception that dance follower bees have of the site location they derive from the dance.

Details About Flights of Recruits to the Goal

Observations with the naked eye help win first impressions. Harald Esch and Joseph Bastian saw the following:

> In two cases we were in the position to follow the arrival of recruits to a feeder over a significant distance. They came directly from the hive in a zig-zag flight at an altitude of about 10 m, down to about 1–2 m and to the feeder [that was scented, J.T.]. [56, p. 180].

Such more or less anecdotal observations of single bees received factual support when it became technically possible to record real measurements of honeybee flight behaviour ([208, 209, 210]; see also Figs. 2.5, 9.2, and 9.3).

Radar tracking of flying foragers allowed for the first time in the research of recruitment behaviour of honeybees, the investigation of events between the hive and the food source. They make a decisive contribution to our conception of how recruits, despite unclear indications in the dance, get to the goal. Old ideas and models can be tested and new questions will arise.

Considering Fig. 9.3, the recorded flight paths at first take a general direction as expected of distant goal orientation (see Fig. 1.1). From von Frisch's discovery, recruits are compass-orientated in this stage (with reference to the position of the sun and/or the polarisation pattern of the sky) and transform the directional information from the observed dance. Termination of flight in a general direction implies

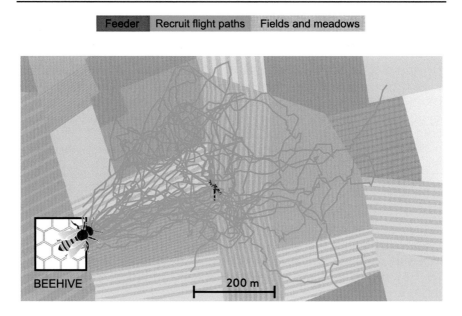

Fig. 9.3 Flight paths of the first outward flight of dance followers [after 168]. For this experiment, the foragers were trained to a feeder. Workers that followed foragers' dances were marked, captured as they left the hive, fitted with a transponder, released, and followed to the goal with radar. These bees first flew from the hive into a sector, with an open angle of about 45 degree of arc

the beginning of the search phase in an area where many cues to the goal converge. Which stimuli the bees finally use to orientate remains an open question.

How, then, does the first section of the outward flight of the dance follower come about? Two hypotheses are proposed: 1. Each dance follower attends a random selection of dances in the hive advertising the food source. In her outward flight, she can exactly follow the information she has observed. The broad spread of flight paths would in this case reflect variation in the dance. 2. The dance follower cannot, in principle, derive exact values (duration and direction of the waggle phase) but instead gains only an approximate indication for the region the dancer advertises. It is, however, irrelevant in both cases for dancers to provide exact or inexact information, or if the dance follower can exactly or inexactly respond. In every case, they can only set out from somewhere within the sector in which the indicated goal lies.

Do Bees Have a Map in Their Heads?

An alternative explanation for how recruits find the goal despite the broad scatter of the initial flight paths does not require communication between bees in the field nor distance orientation. It is suggested that bees have access to a map of the landscape in their heads, created during their orientation flights. In relation to these maps, they apply vector calculations and derive the correct vector flights that bring them to the goal [170]. The dances supply the values for the calculations.

The idea of cognitive maps enabling the orientation of bees in the field is disputed. One finds some research studies that strongly doubt this [e.g. 30, 37, 48, 288], and some that argue for this model [e.g. 28, 29, 170, 171].

An experiment that at first sight appeared to support the idea of cognitive maps led to a contradictory conclusion when repeated: Bees were trained by the first research group to a feeder set up on a boat anchored in a lake [100–102]. The observations spoke for the idea of a cognitive map [100], because no recruits set out for this remote goal and no newcomers were observed although dancers trained to the site advertised it. The interpretation: Dance followers already knew from following the dance that the advertised goal lay in the middle of a lake, an "impossible site" for which one did not even need to set out. A similarly arranged experiment of a second research group could neither confirm this conclusion [288] nor support the idea of a cognitive map.

Given, as in this case, several adequate possible explanations exist, the simplest of them is to be preferred. Johannes Clauberg (1622–1665) wrote the following about this tried and tested principle, known as Occam's razor: An explanation is simple when it contains as few hypotheses as possible. Under this aspect, the two hypotheses are presented here (distance orientation or map). The first is the more parsimonious to account for the fan-shaped distribution of flight paths: On the first section of their outward flight, dance followers cover a wide area indicated in the dance and reach the goal after a search phase with the aid of an additional goal orienting stimuli.

How large the Phase 2 search area of distant goal orientation can be has not been investigated. One can suppose the motivation of recruits to play a role, for how long and extensively they search to find goal orienting stimuli.

How Many Directional Information Bits Are in a Bee Dance?

Information transfer is central to communication. The more accurate the details, the greater the information content. The precise locational information in a bee dance is emphasised in countless publications (see also Fig. 3.6). Merely, a single example is given here: The well-known nature photographer, Sir David Attenborough, commented in a three-minute video from the Smithsonian, a world-famous museum complex, about the information transfer through the waggle dance as follows:

[…] her instructions are remarkably accurate and can pinpoint the location of a nectar source over 6 kilometres away,

How extensive must the information content be for such exact site details? How large, actually, is the information content of a waggle dance?

The most frequently applied definition of information asserts that information reduces the uncertainty about the facts of a case [239]. Bee dances reduce the infinite possible locations of sites surrounding the hive to that for which a dancer advertises. The information content transmitted in a dance can be formally expressed in a unit of measurement, a bit (Binary unit). A bit is defined in terms of the information content of just two possibilities. An example: If one is able to choose only between black and white, the indication of one of the two is exactly one bit of information.

Applying quantitative information theory to the bee dance is extremely helpful in estimating the accuracy and the importance of the dance. The contribution of a bee dance to the discovery of an advertised goal can be estimated from an information theoretical viewpoint. The higher the information content in a dance, the less uncertainty about where a dance follower should fly.

Should a bee at the start of a foraging flight choose between two sectors, equally distributed in two general directions (for example, the choice is either North or South), the information amount sending the bee in the right direction is exactly one bit. If four sectors are available with four directions (North, South, East, and West), she must choose one from four directions and consequently the information content for a goal orientation message is now two bits. For a goal within an angle of ten degrees of arc, there are 36 possible sectors. This means one from 36, requiring a message information content of 5.2 bits. Were the angular instructions in a dance down to one degree (360 possibilities in a complete circle), the corresponding information content of the message giving the correct direction (one out of 360) is exactly 8 bits.

The more imprecise the direction, the less the information content in the message and, vice versa, the higher the information content in a dance, expressed in bits, the more precisely an intended direction is indicated, which is intuitively clear without the mathematics.

Are these considerations far too theoretical or do they bring us further in investigating the language of bees? From von Frisch's publications, John Burdon Sanderson Haldane and Helen Spurway [113] calculated the directional indication in the waggle dance to have an information value of 2 bits. Such dance information would theoretically limit the directional choice of a dance follower to be between four sectors each with an open angle of 90 degree of arc. For their outward flight, bees must choose between four sectors each of ninety degrees.

The information content in dance movements calculated by Roger Schürch and Francis Ratnieks [225] came from their own measurements. They reckoned an information content of 2.9 bits, which means the sector in which the dance followers were sent had an open angle of 45 degree of arc.

These calculations of the open angles of sectors illustrate the accuracy of the dance. A recruit that begins her first outward flight in the direction of the advertised goal has, theoretically, for a transmitted information content of 2.9 bits, eight sectors each with an open angle of 45° to choose from. Based on an information theory view, the dance in this case cannot provide a direction with an error of less than 45°. Were the entire information content of 2.9 bits of the dance employed, the open angle of the sector into which the recruits fly amounts to about 45 degree of arc.

These statements can be tested by recording exactly how the dance followers interpret the directional information from the dance in their outward flights. Such a test was undertaken for the first time for this book.

Radar tracking of flying bees provides a technique for checking the calculations of Schürch and Ratnieks. A publication appeared in 2019 in which a number of flight paths of dance followers were illustrated [168]. The flight paths were distributed over a sector that appears to match the open angle of 45°, as estimated by Schürch and Ratnieks.

The experimentally determined sector covered by outward flight paths (Fig. 9.3) could not have been narrower because the information content in the dance was not large enough. However, it could have been broader. Nevertheless, it is not. Its dimensions correspond to the information given in the dance and the bees do not ignore this.

This view spins a thread further in dance language research. With his pioneer achievements, von Frisch provided the initiative. Gould could show qualitatively that the dance exerted an influence on how the newcomers distributed themselves in the field. The convergence presented here of applying information theory to the bee dance and recording flight paths brings a quantitative aspect into the picture. It shows that the transmitted information content in a dance about flight direction is fully employed.

Karl von Frisch chose angular differences for his fan experiments of eight, ten, and fifteen degrees between neighbouring control stations (see Fig. 3.5). To communicate reliably in a sector with an open angle of eight degrees, the theoretical information content required is 5.2 bits. For a fifteen degree open angle it amounts to 4.5 bits, significantly more than the dancer offers [113]. No wonder recruits spread themselves over an area broader than the steps between the neighbouring control stations. The spread of the recruits is even clearer when the control stations lie closer together (see Figs. 5.5 and 5.6).

In Gould's experiments, modelled on those of von Frisch, the angular separation between every two stations was only three degrees (see Figs. 5.5 and 5.6). Were such an angle correctly communicated, the information content of the waggle dance would be 6.9 bits. This value lies far above the maximum possible information content calculated for the orientation of a waggle phase to indicate a direction. Based on this analysis, one can reiterate the suspicion that experimentally discovered distribution of recruits over the control stations was only partially influenced by the dances. The situation and events in the field, such as the flight traffic of the bees trained to the feeder, and the scent of the control stations, are additional sources of information for the newcomers.

How Many Distance Information Bits Are in a Bee Dance?

The information content of a dance related to the distance to a goal is, according to Schürch and Ratnieks, 4.5 bits. In their calculations, they used the duration of the waggle dance phase (data from 35) and took a maximal flight distance of fourteen kilometres [225]. For a flight distance of only half of this much (seven kilometres), the information content of the dance amounted to 3.6 bits.

It is somewhat more difficult to clarify the value of the information content for distance than for direction. In principle, this means that for information content of 3.6 bits the stretch between the hive and the feeder is divided into twelve equal sections and the distance to the goal cannot be more accurately indicated than the length of one of these sections. For a distance of seven kilometres between the hive and the goal, this means the bee is sent to a zone from 600 m behind to 600 m in front of the goal.

The closer the feeder lies to the hive, the less the calculated information content in the dance and the less accurate the site description, a fact that occurs to one by simply watching the dance (see Fig. 7.4).

The thoughts and observations set out here say nothing about what value (i.e. how useful) the information mediated in dances has for the colony in general. They concern only the communication between dancer and dance follower. The value for the colony as a whole essentially depends on the size of the colony [44], and above all on what flowers are offering, available within the colony foraging range [43, 148, 193]. If a flowering tree is the only nectar source in the flight range of a colony, a message with information content of only 2.9 bits is already very helpful for foragers. On the other hand, if bees find nectar-rich flowers all around the hive, such information at the level of the colony is worth much less or even irrelevant [148].

An Interim Balance Sheet

10

Misunderstanding, Blind Spots, and Expedient Hypotheses

Previous research on the dance language of bees concentrated mostly on what happens in hives and at food sites and ignored events in the field. The classical model (Fig. 3.6) has consequently taken on the status of an icon.

As presented in this book, information flow between bees in the dance mediates half-true indications to dance followers about the location of a single goal point. In fact—as the provided arguments show—this is irrelevant for the bees. Instead, the dance follower is sent only to an area by the dance (the first phase in distant goal orientation), where she then searches (second phase) and from where she finally is directly attracted and/or led to the goal (third phase).

Virtually, nothing is known about the behaviour of bees in the field. Nevertheless, many expedient hypotheses support the classical model of the bee dance.

Researchers Have More Data Than the Bees

Researchers analysing the dance have a complete overview of all available information from a dance advertising a goal. A dance follower does not. In researching the recruiting mechanism, it helps to think like a bee.

Researchers of the dance language can derive information from a bee dance about a site's exact location, if they employ the appropriate technical effort, collect sufficient data, and dissect the dance into single separate fragments. Such data are more accurate, the greater the number of dance rounds analysed and the extent of the sample from the total of all previous dances. Prepared and derived this way, the data permits the statement that the bee dance contains information about the exact location of the site (see Fig. 3.6).

J. Tautz, *Communication Between Honeybees*,
https://doi.org/10.1007/978-3-030-99484-6_10

This high information content, calculated purely by human observers, is believed to represent what is exchanged in a dance between the sender (dancer) and receiver (dance follower).

This assumption is a misunderstanding because it overlooks the fact that a dance follower receives only a fraction of the data available to researchers about the dance. The scientific view is a derived concept, an abstract result of the summary, and mathematical treatment of many dances. The actual information flow in the every day of honeybees, offered by the dancers, is something very different. Just how little information exchange takes place is quantified at the end of Chap. 9

An Example: Understanding the Orientation of Homing Pigeons

Research on homing pigeons provides a good example of careful and simultaneously conceptually determined studies in a different area of behavioural biology, with a similar database. For pigeons, transported and released hundreds of kilometres away from their home roosts, one can determine the initial flight direction compared with the direct course to the goal, their home roost. The accuracy of an averaged vector, calculated from the initial directions the pigeons take, increases with the number of observed values (Fig. 10.1).

However, researchers did not conclude that pigeons, at the start of their flight home, knew exactly the correct path to the goal because the mean of the flight directions corresponded with the direct path to it. The broad spread of observed flight directions would not support such a conclusion. Instead, observations and experiments showed pigeons reached their goal through an initial gross orientation that became increasingly precise during the flight [32], as expected in distant goal orientation.

The similarity between the sequence of goal-orientated phases in the homeward flight of pigeons and the flights of bee recruits to a goal is obvious. At the start, the animals are without information about the location of the goal and cannot fly directly to it. However, after searching at a certain distance from the start, they come across cues that lead them directly to the final goal. This is a distant goal orientation.

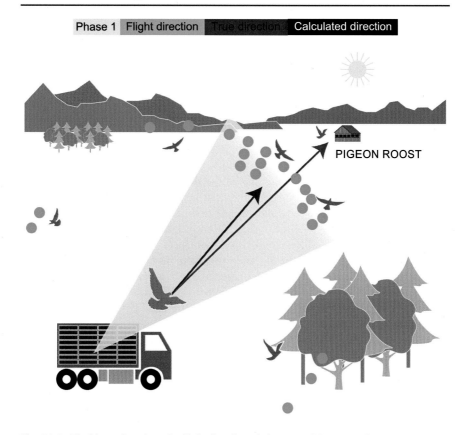

Fig. 10.1 The blue points show the flight direction of pigeons 1,500 m away from the start. They are on their way home, twenty kilometres away. The statistically calculated mean of the measured flight directions of the individual pigeons (black arrow) almost reflects the direct path to the home roost (red arrow), where the pigeons finally arrive. Here, Phase 1 of distant goal orientation is demonstrated

Research on the Dance Ignores Behaviour of Bees in the Field

The central issue of the dance language is: The dance provides honeybees with the direction and the distance to a goal.

Another misunderstanding focusing research on the dance language comes from calculating means that arrive at a precision for a goal instruction in the dance (see Fig. 8.1) and taken to be the actual information content in dance. Observation of exact arrivals of recruits at advertised sites is believed to confirm this view. Consequently, there was no need to include any consideration of communication between bees in the field in concepts and research programs.

Over decades, countless experiments carried out and published have one aspect in common: Apart from very few exceptions, all interpretations of observations and data about recruitment in honeybees, gained in the hive and at the feeders, are entirely without any consideration of communication of honeybees in the field. How are the complete flight patterns, directions, changes of flight paths, distances, flight velocities, etc. influenced by a previously observed dance? We know very little about these.

The presence of experienced bees, and orientation signals and cues from the goal, must be absent if we are to distinguish between the contribution of gross instructions in a dance and of communication in the field, in discovering the goal. The sole effect of the dance on flight patterns of dance followers can be investigated only under such conditions. Separating the various influences (dances in the hive, communication in the field, and the scent of flowers) is an almost unsolvable task for an experimental resolution of these different influential factors.

So far, in the extensive literature on bees, there have been only two experimental attempts to study the behaviour of dance followers on their outward flights to an area indicated in the dance that had not been visited by other foragers and where no feeder was present. These experiments concentrated on the first phase of distant goal orientation. The third phase of distant goal orientation, guidance to the goal, was absent (see Fig. 8.2).

Just Two Experiments Focus on Pure Dance Information in the Field

In the first of these studies, the so-called tunnel experiments (see Fig. 7.5), dancers advertise a non-existent food source. They indicate a far longer flight path than the actual distance between the hive and the feeder [59]. Slowly flying recruits are clearly observable and easily captured in a net. In this experiment, observers positioned at five separate stations captured the highest number of slowly flying bees around the site advertised by the dancer as the goal [59]. The study was not concerned with where the captured bees switched to the Phase 2 search phase (see Fig. 8.2) or how temporally and spatially extended these search flights were.

However, dance followers clearly estimated their distance to the feeder from the half-true information in the dance. The transition to slow flight, allowing the capture of bees, suggests that in their distant goal orientation, Phase 1 terminates at a reasonable distance from the goal and merges into Phase 2 even without the presence of experienced bees visiting the site or the existence of odour.

A different experiment carried out by Joe Riley and co-workers [209] excluded the presence of experienced bees near the goal by capturing the dance followers after they had followed a dance and attempted to leave the hive. Recruits were fitted with antennae allowing them to be tracked by radar, transported to a distant site and released, tracked, and their flight paths recorded. The flight paths were not randomly directed. Each bee flew in a general direction confined within a sector. The open angle of these sectors, estimated from the published data, is thirty to forty degrees of arc. Here, clearly, the dance followers employed the half-true directional information from the dance.

Areas of the sectors covered by flight paths correspond to the theoretical information content of about 3 bits in the directional instructions of the dance. A similar information content for the dance was also obtained when calculated in a different way [225; see p. 168ff].

In Riley's experiment, the distance to the feeder indicated in the dance was 200 m. The recorded flight paths in the publication show a flight distance of about 500 m but no noticeable events at 200 m, the goal distance. It also remained unclear from the publication how far bees continued in the same general direction, or at what distance they terminated the first phase of distant goal orientation to begin the second phase and their search for experienced bees or flower scents.

Both experimental approaches have the potential to show how widely spread the area could be in which bees begin to search for goal orientation stimuli and exactly where these lie. In some cases, the search area may turn out to be small should, for example, searching bees immediately come across experienced bees on their way to the goal, or a wind bearing scent from the goal directly towards the outward flight sector. Or it could be correspondingly large when such stimuli were not initially detected.

Newcomers unable to find the advertised goal and returning to the hive after many minutes to half an hour could speak for an increasing spread of the search area. It could also imply a continuous patrol of the same area, or a combination of both. Such questions remain open.

The half-true statement of dances imposes the impossibility of bringing recruits to the goal exclusively with the dance. However, the question is whether highly accurate dances giving reliable site co-ordinates without variability would bring recruits directly to the goal. A three-phase, distant goal orientation would then not be involved. The employment of mechanical dancers with precise movements (see Chap. 9) could exclude unclear dance information. An appropriate experiment, though, could only be carried out if no other bees are present in the field and no stimuli stemming from the feeder, a significant challenge and perhaps a virtually unsolvable project for future research.

Expedient Hypotheses and Assumptions Support the Popular Model of the Bee Dance

The conclusions from many experiments and observations, derived from an uncontested underlying data set, are sufficient for the meaning of the bee dance related to the recruit's search for a food site to be re-considered. Nevertheless, based on additional assumptions, the data have been adapted to fit the classical model (see Fig. 3.6) so that a new interpretation of older ideas has not been considered necessary.

Four broadly formulated examples:

1. Observation: The waggle dance provides inaccurate information about the site.

Expedient hypothesis: The problem is indeed contained in not only the dance but also the solution. Dance followers derive accurate information from the inaccurate dances, while they are able to determine the location of the site by calculating the mean of many dances.

2. Observation: The waggle dance provides inaccurate information about the site.

Expedient hypothesis: If bees do not calculate average values of site information, and so far it has not been shown that they do, this should not be a problem for bees because seldom is there only a single flower at the site advertised by the returning dancer. If bees distribute themselves over a wide area as indicated in the dance, they would visit more flowers—the inaccuracy of the dance is to the advantage of the colony.

3. Observation: Dances bring recruits near to (what does "near" mean?) advertised flowers, but not without scent.

Expedient hypothesis: The dance information brings the newcomers to the goal where the scent of the flowers motivates them to land. They reach the goal without any further assistance but return to the hive without landing if the site is not scented.

4. Observation: Recruits find a feeder without any useful dance information, from the scent alone.

Expedient hypothesis: The dance is one orientation guide from several equal alternatives available to the recruits. Bees can decide between assistant stimuli and exclude other possibilities when the dance is unhelpful. There are also occasions when the dance alone brings the recruits to the goal.

In these four examples, and without summarising and ordering other provisional assumptions about observations of the bee dance on the one hand, and on the other, what is expected from it, communication between foragers in the field plays absolutely no role. The dance remains in spite of all, the only communicational bridge between bees.

We Know Little About Communication Between Honeybees Outside Their Hives

Honeybees live in two worlds, in their dark hives and the bright world outside. They spend the greater part of their lives in the hive and only leave it as older workers for foraging flights, if one excludes annual swarms when about half the colony takes part.

The emphasis of bee research so far has treated honeybees to have dual personalities. In the hive, they behave as social insects but underway in the field as solitary individuals that at most leave a scent signal on flowers they visit.

This emphasis is abundantly clear after a single glance at scientific publications on this theme: More than a thousand studies on the behaviour of bees in the hive set against a handful of works devoted to observations of communication between workers on their flight to the goal. Most other studies about signals of worker bees in the field describe chemical marking of flowers indicating to following foragers that the nectar store is depleted [94, 183], or acting as an attractant for newcomers [158].

The dance sends recruits into a sector in the field where the search phase for goal orientation signals begins, finally reaching the advertised site. The blind spot concerns the last two phases of distant goal orientation in honeybees.

We know nothing about the intermediate Phase 2, the search phase, apart from that it exists [68]. There is evidence for a lateral spread of the search phase, supported by the time foragers are absent from the hive. One measures how long a dance follower, after a dance, is underway and returns without nectar or pollen. Not much more than this is known about the search phase. We do not know how widely spread the search area can become, or if it is determined, if it varies, and what form it takes depending on the conditions.

One can also only speculate over the last phase, the near goal orientation Phase 3. The thoughts and statements found in the literature are correspondingly vague. Karl von Frisch, in his early studies [72], at first concerned himself with this phase, but did not follow it further. Adrian Wenner and his co-workers contributed a great deal of basic data relevant to Phase 3 of distant goal orientation, in particular relating to the importance of scent coming from the site for the goal-seeking recruits [284]. Communication between bees was not mentioned in the publication.

The sparingly described end phase of the goal-finding honeybee recruits stands in stark contrast to the many details known about Phase 1, beginning with the highly detailed studies of the dance of bees. The brief accounts on Phases 2 and 3 in terms of explanation content are as though one described the entire knowledge about Phase 1 merely as "At the start, foragers are sent by the dancer into the field".

Unequal Technical Chances

Considering the actual investigative possibilities and working conditions offered to bee researchers, the imbalance between studies of bee behaviour in the hive and out in the field reflects what confronts them in the two worlds of bees.

Fig. 10.2   An observation hive, as it has stood for a hundred years in the centre for research on the bee language

If one establishes a bee colony in an observation hive (Fig. 10.2), with combs arranged in two dimensions, no detail of the bee behaviour can escape the eye. Direct observation, analyses of video recordings or with the help of other technical systems could capture every interaction between all the bees.

These ideal methodological conditions make it clear that the research on the communication of bees looks for answers in the hive when the secrets in the field are hidden (one is reminded of the old joke cited at the start: One searches where there is light). A study of communication between honeybees in the field is, in its present technical application, far short of the possibilities offered to researchers by an observation hive (Fig. 10.3).

Fig. 10.3 Honeybees (here two bees on their way to a feeder) fly up to thirty kilometres per hour over a distance of many kilometres including through vegetation. The observation of freely flying bees in the open and detecting signals they leave behind them is extremely difficult. It is not yet possible when they pass through natural vegetation

Swarm Behaviour Shows Bees Communicate Not Only Through Dances

11

The dance communication may have its origin in swarm behaviour [103] where the inadequacy of the dance alone is particularly clear. Instead, it is a single building block in an unbroken behavioural chain of goal orientation events. Recruitment to a feeder may represent mini-swarm behaviour because it employs the same behavioural building blocks.

Dances are helpful for the day-to-day business of a bee colony, harvesting nectar and pollen, but dispensable and at times it is even better without them [148]. This does not apply to swarming. The overwhelming role played by communication between bees for the existence and survival of a bee colony is obvious in bee swarming.

The bee dance is essential when scout bees have discovered a new home and need to gain support for their discovery. However, it also becomes clear that half-truths in the bee dance would mean the death sentence for the colony were the inadequacies in the dance not complemented by other forms of communication.

A swarm of 20,000 or more bees that leave their old quarters with the old queen, and set up a temporary bivouac outside the hive, must quickly find an appropriate new home. All the bees must finally make their way to a new nest site, often in a hollow tree trunk in a forest habitat.

We know many details about swarm preparation and the behaviour bringing swarms to a goal from the contributions of Martin Lindauer, the first to concentrate on dances on a swarm cluster [150] and Thomas Seeley [231].

How Single Bees Advertise New Homes

A critical interaction between a relatively small number of bees takes place before the entire colony is included in the communication. Failure at this stage would mean an early end to the swarm.

A swarm, once it leaves the hive, has to act promptly. Provisions for the journey, the full honey stomachs of workers, are limited and rainstorms can seriously damage an unsheltered colony hanging from a tree branch. There is also competition for the best sites, should several other colonies in the area swarm at the same time. Two to three hundred scout bees start out from the swarm cluster, settled not far from the old hive, to search for a new home (Fig. 11.1). Scout bees are usually older foragers that switch from searching for new food sites to the search for a new home. Not all old foragers are able to make this transition. As the development of different characteristics and abilities of individual bees, genetic make-up is the decisive factor. The likelihood that a worker will be a scout bee depends on genes she has inherited from her father [212].

If it ended there, with each scout discovering a new nest site and not able to pass this knowledge on to other bees, the swarm would be lost. Every bee in the swarm must reach the same goal (Fig. 11.2).

Decisively important is that each scout bee succeeds in sharing her discovery with as many others as possible, to convey its nature and to promote it with a dance. Mini-swarms, or even a single newly won bee, must be brought to the new site. The

Fig. 11.1 The "promotion" of a new home proceeds step by step. Here, a scout bee has brought the first new bee from the swarm cluster to this tree. Bee by bee, a small group develops that is familiar with the site

Fig. 11.2 Scout bees that have found a new nest hollow advertise it with their dances on the surface of the swarm cluster, on the bodies of other bees. These dances reach only a small fraction of the bees in the swarm and are not adequate for recruitment to the new nest site—only an unbroken communication chain of various stimuli can succeed in bringing the entire swarm to their new home

assembly of a small cohort for the new site proceeds slowly, bee by bee. On average, a new recruit joins every five minutes [236], until at the end twenty or thirty bees (in some cases up to fifty bees) belong to the group [231].

Additional scouts are each individually recruited by waggle dances on the swarm surface and will later dance themselves and advertise the new discovery.

The angular divergence of sequential waggle dance rounds (see Fig. 7.4) in dances on the swarm cluster over the bodies of other bees is somewhat less than in dances on the comb [281]. Contrary to what was at first thought, bees do not have different dances for food sources and new nesting sites. The different physical conditions of the substrate (comb or bodies) imposed on bee movements determine their dances and the angular divergence between dance rounds [251].

Impossible Without Communication in the Field

Everything that research has presented so far to confirm the classical thesis that despite inaccuracies the bee dance leads the bees directly to the food sources falls away in the recruitment of small cohorts to a newly discovered nest site:

1. In this most critical phase, bees cannot derive an averaged correct instruction from many (half-true) dance rounds because at first only a single bee dances for the goal.
2. To spread out over an area, advantageous for a broad field of flowers, would be fatal. A single advertised hollow must be visited, virtually a single point in the landscape.
3. No goal orienting signals or cues emanate from the hollow.
4. Without a continuation of communication in the field, each scout bee would remain alone with her discovery and no recruit would arrive at the new site.

In this decisive stage in the life of a bee colony, the allocation of the dominant role to the dance language cannot function. The continuation of communication between bees in the field when swarming delivers a decisive contribution to goal orientation.

In their efforts to win new followers to the newly discovered nest site, scout bees employ the same behaviour known from recruitment to food sources: they dance, fly back and forth between the swarm and the goal and at the goal, perform buzzing flights and scent marking with everted Nasanov glands (Fig. 11.3).

Fig. 11.3 A forager that had previously danced in the hive, arriving at flowers for which she had advertised, with a visibly everted Nasanov gland

We know nothing about what passes between scout bees on their way between the temporary swarm bivouac and the new home. This gap in our knowledge is a blind spot, like that in recruitment to a food source.

In contrast, we know much more about the flight of the entire swarm to the new nest site [231].

How the Swarm Reaches the Goal Without Dance Information

Every scout bee that has discovered a new nest site dances for this on the swarm cluster, as do all the others that she had taken to the site to inspect. In a voting process between the scouts, which Lindauer first explained [150], only the best of all potential sites finally remains and is promoted in all waggle dances [233]. Fifty or more scouts recruited to this site now dance exclusively for it on the swarm cluster surface. The recruiting process that will guide the entire swarm to the new nest site, often several kilometres away, now begins.

The majority of bees in a swarm cluster know nothing about the approaching departure. More than 95% of swarm bees are unaware of the dances and have no idea about when a flight will begin or where it will lead. Maximally, only five per cent of the swarm have visited the goal before the swarm sets out for it [231]. The bee dance as a directional indicator for moving the swarm from its temporary bivouac to the new quarters falls away completely.

How do the few bees manage to guide the entire swarm to the goal? The complex behavioural chain of events plays out in these steps: To start, the swarm is brought into a departure state and "flight motors" warmed up. Thereafter, a signal is given to take off. Once in the air, the swarm is led to the goal. The same bees that advertised the new site as the best possible accommodation with their dances are also responsible for these three steps.

About one to one and a half hours before the swarm takes off, high-frequency "piping" tones occur in the swarm cluster [219, 220], like those that also occur before bees swarm out of their old hive [197]. Scout bees in a swarm cluster generate the piping tones. Once a decision is reached about the best site, scout bees first fly back and forth between the swarm cluster and the new accommodation, perform waggle dances on the surface of the cluster, and mark the goal with scent. The initial small group of scout bees familiar with the site increases in size.

After a certain time, scout bees extend their behavioural repertoire. They break off their waggle dances and begin to run rapidly over the bodies of swarm bees, pause, grab onto another bee with their legs, and emit a piping tone generated by their flight muscles. Soon after beginning this new behaviour pattern on the cluster surface, they vanish down into it. There they energetically continue piping [232]. Because piping bees within the cluster have to force their way through the closely packed colleagues, the vibratory pulses of the piping bees quickly reach a large number of workers as they make their way through the cluster. The "piped" bees react to the piping stimulus by raising their body temperatures. Piping intensity increases until it is almost continuous, about thirty minutes before take-off.

The overall temperature of the cluster rises. When the temperature of all bees has reached between 35 and 37 degrees Celsius, the swarm is ready to go [232, 237] and waits for the signal to start. The actual signal for take-off involves a so-called "buzz run" [150, 211]. Shortly before the start, more and more bees run hectically over the cluster. These are the same bees that piped, that are familiar with the new nest site, and had previously performed the dances advertising them [231].

The Swarm Underway

The swarm take-off is an explosion. The bees fly off together. They do not peel off the cluster in layers. The swarm first hangs like a large living balloon in the air above their temporary stopover. Slowly, it changes form by stretching out lengthwise to take on the shape of a zeppelin. This begins to move slowly and then with increasing velocity until it reaches a maximum of eight kilometres per hour, directed exactly towards the new home.

Guidance is by the relatively small group of bees that have already carried out all the steps from the discovery, inspection, and promotion of the new nest site and the swarm take-off.

Three sources of stimuli hold the swarm together and bring it forward. A mix of pheromones with 24 identified components and produced in glands in the head of the queen achieves cohesion of the swarm. In addition, the queen possesses so-called tergal pocket glands in every abdominal segment. Secretions from these also hold the workers together. Honeybees are highly sensitive to their species-specific pheromones and can detect very small amounts released into the air.

The pheromone from the Nasanov gland acts as an attractant in the swarm. Geraniol, an important component of the pheromone, is produced in the Nasanov glands of workers (Fig. 11.3). Alphonse Avitabile and his co-workers describe [7] that above all, pheromones guide a swarm to the goal and geraniol is of particular importance. Scout bees scent the new site in buzzing flights and its entrance by walking around and marking it with this pheromone.

Geraniol release is assumed to occur when bees flex the joint between the last two segments of their abdomens, exposing a visible pocket into which hundreds of gland cells secrete liquid geraniol. The anatomy of the Nasanov organ and physio-chemical properties of geraniol suggests geraniol is released as soon as it is synthesised and due to the low surface tension of the cuticle, moistens the outer surface of the bee's body although the organ cannot be seen with the eye to be everted (Fig. 11.4).

Fig. 11.4 The Nasanov gland is situated in the joint between the last two abdominal segments (arrows) and takes up the entire breadth of the gap between the segments. The last abdominal segment is covered with more hairs than that anterior to it. The liquid pheromone may be drawn out through the gap between the segments, even if the gland is not visibly everted, and held by the hairs—although so far this is still unclear

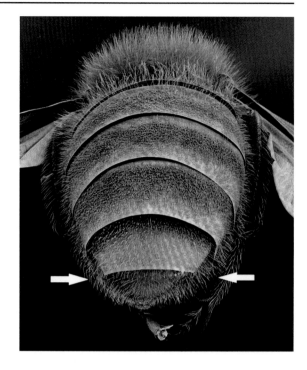

In addition to chemical signals, optical signals for orientation also come into consideration. Lindauer observed single bees in a swarm that flew with high velocity straight through the swarm. The axes of their flight paths lay in the direction of the goal [106, 150].

To test the relative importance of chemical and optical signals in guiding a bee swarm, Madeleine Beekman and co-workers covered the gap between the last two segments of the abdomens of all workers in several small swarms, with a dab of paint, thus blocking the Nasanov organ. Despite this, all swarms reached their goals [14]. The conclusion of the author that the swarms are not led by geraniol, although it plays a role in every case where worker bees are attracted to a site, will certainly be re-evaluated when there is evidence that the bees in her experiment were unable to release the pheromone.

Rethinking Communication Between Bees

12

It can be established, from an analysis of the relevant publications (consulted here) that build on von Frisch's epochal discoveries, and the in-depth re-investigations of his followers, that: The dance information is related to Phase 1 of distant goal orientation, no more and no less. However, considering Phase 2 and Phase 3, one comes across the repeatedly mentioned blind spot. How can we close this?

A Discovery Astonishing as the Dance of the Bees

It may lie with the methodological problems associated with bee research in the field that one of the most exciting discoveries has attracted almost no interest and gained no scientific attention. Karl von Frisch [78], John B. Free, and Ingrid Williams [67] were the first to describe, about 50 years ago, the connection between the communication behaviour of experienced bees at a goal site and the site's own particular odour. Characteristics of a goal can influence not only the dance in the hive, the dance language, but also communication in the field.

The less (for human) perceivable the odour of a natural or artificial food site or water source, the more noticeable (for human observers) the employment of bees' Nasanov glands.

The same behaviour appears in swarms when scout bees scent the entrance to the new nest site that emits no natural attractants for bees [66, 67]. Previous comments on the research of this behaviour are purely descriptive and present no measurements or data.

These observations lead to an astonishing interpretation: Experienced bees complement naturally occurring olfactory stimuli that flow from a goal with their own signals, closing the communication chain that began with the dance in the hive. The fact that this complementation is graded and apparently adapted to need is amazing and, in importance and accredited performance of bees, can be set at a level equivalent to the start of the communication chain in the hive in the form of the dance.

© The Author(s), under exclusive license to Springer Nature Switzerland AG 2022
J. Tautz, *Communication Between Honeybees*,
https://doi.org/10.1007/978-3-030-99484-6_12

How do experienced bees that had previously danced in the hive "know" that searching newcomers are underway in the field? How do they know that the site they advertised in their dance emits weak, or no orientation, signals? How do they know to what extent they need to assist the searching foragers in the field?

Questions upon questions...

What Role Does Scent Play in Bee Communication?

The decisive supporting role experienced bees play for inexperienced bees in the field was plain to Karl von Frisch and other early bee researchers like Bruce Lineburg [155]. They were also aware of the consequent importance of the Nasanov gland pheromone.

This gland is located on the dorsal side of a bee's abdomen between the two last segments (see Fig. 11.4). Its discoverer after whom the organ is named [291] had no idea about the function of the organ, first recognised in 1900 by Fredric William Lambert Sladen [242] to be a gland that produced a mixture of odours, with geraniol as one of the main components.

The problem at the time: There was no known way to measure the concentration of geraniol and its spatio-temporal distribution along the bee flight paths. One depended on one's own sense of smell.

Von Frisch wrote:

[...] with the wind in one's face, one noticed the characteristic scent when they swarmed around the feeder with the organ extended. At a distance of at most 1 to 2 paces, I could no longer detect the odour. For the bees this cannot be the case because the stream of arriving newcomers of a richly fed swarm is so strong that apparently searchers flying around from an influenced area of very different dimensions are attracted. [72, p. 171].

The absence of appropriate research methods is regretted:

The olfactory organ of the bees for this attractant (odour) that they produce themselves is particularly finely tuned. I know of no way to find an exact measure for this. [72, p. 172].

The absence of knowledge at the time was substituted by assumptions and there have been, to this day, no thorough investigations. The original assumption of von Frisch related to the release of the pheromone from the Nasanov glands still stands:

If, UNDER PARTICULARLY FAVOURABLE CONDITIONS, scent marking the way during flight is effective, remains an interesting question, as long as we do not know if the foragers fly along the stretch with extended scent organs. I believe they first make use of these when arriving at the goal. [78, p. 229, emphasis in the original].

High-speed video recordings can capture an objective view of a maximally open gland of slowly flying honeybees, so far only observed by the eye. A new, still unpublished study using such high-speed video recordings shows the last abdominal segment posture, and so Nasanov gland opening, to depend on the bee's motivation, that is, if it had danced in the hive or not. While gland-opening stages are difficult for humans to judge by direct observation, with videos a systematic study of a bee's arrival behaviour at a feeder is now possible (Benjamin Rutsch-mann, personal communication).

Stretches of the flight to a goal, over which experienced bees that danced in the hive open their Nasanov glands, remain unknown. It would also be good to know if geraniol is released from the gap between the two last abdominal segments only when the gland is recognisably open, or if the liquid pheromone has already been drawn out over the last abdominal segment by capillary action. The last segment is more densely covered with hairs than that anterior to it (see Fig. 11.4). A thin secreted film could adhere to the hairs and evaporate from there. To speculate a little further: The conspicuously large number of fat cells in the body cavity housing the Nasanov gland could provide a fixative promoting a gradual release of the pheromone.

The exploration and testing of these speculations are possible with currently available methods.

How Sensitive Are Bees to Odours?

Adding to some unfortunate realities about bee communication research, there is not a single (!) physiological study of worker bees' sensitivity to the important Nasanov gland communication pheromone.

The sensitivity of bee antenna receptor cells and brain nerve cells to geraniol and other components of the Nasanov gland pheromone is not known.

Workers at the entrance to a hive use scent from this gland as an attractant; it plays a decisive role in swarming, and Karl von Frisch, before his discovery of the dance language, regarded it to be responsible for recruitment to a goal site.

In his efforts to penetrate the communicative world of bees outside the hive, von Frisch used a comparison with the human olfactory sense to gain insights into the bees' olfactory world. He established the detection threshold for two substances (bromstyrol and methyl heptenone in liquid paraffin) for himself and his wife and took the results as the baseline for corresponding thresholds for bees [70]. Without instruments to measure scent concentration in air, a series of dilutions was prepared, and the dilution level was recorded that released no observable reaction from bees (and people). Von Frisch found the threshold sensitivity of bees to methyl hep-tenone to be a dilution factor of one in 2,000. Based on these results, he assumed that honeybees did not have a particularly high sensitivity to odours.

Charles Ronald Ribbands [206] found a threshold dilution value, for the same substance, of one in forty million in honeybee learning experiments. Bees are twenty thousand times more sensitive than von Frisch thought. Experiments with bees trained to visit sites with certain odours revealed significantly higher sensi-tivities [207]. As a result, Ribbands estimated the sensitivity of bees to odours to be correspondingly high.

Rodrigo De Marco obtained a highly interesting, and so far unpublished result and contributed it to this book. It extends von Frisch's discovery [72] of the close connection between waggle dances in the hive and buzzing flights heard and seen at the feeder. De Marco determined what von Frisch had already seen but not measured, namely the direct relationship between the duration of a forager's dance and how long the same bee's Nasanov gland was open during its buzzing flight at the feeder and after landing. The longer a bee dances in the hive, the longer the buzzing flight duration and period after landing with a visibly open Nasanov gland (De Marco, personal communication).

Observed signals sent out by experienced bees in the recruitment of bees, both in the hive and at the feeder, are all equally intense. Experienced bees communicate at all known levels from the hive to the goal with equal intensity.

Tracing Odours in the Field

It is hard to imagine that active enticement in the field by experienced bees begins only when human observers notice it. How these bees have behaved before they visibly arrive at the feeder is unknown. Recording the flight paths of bees could contribute a significant step in closing this gap.

Simultaneous radar tracking of many flying bees (see 208 for method) has succeeded with foragers trained to a feeder with geraniol [169]. A similar experiment for the future could track experienced and inexperienced bees simultaneously and would clarify where between hive and feeder the arriving group formed, examine the assumption about the origin of flight curves (see Fig. 9.2), and a meeting of experienced and inexperienced bees.

Assembly of recruits and experienced bees in flight could be based on olfactory and/or optical signs and signals.

Olfactory signals, pheromones, belong to the most important information media in the insect world. Insects not living in colonies use pheromones as sexual attractants to bring males and females together, and a particularly large number of studies on the biology of sexual attractants have been made on butterflies and moths [4, 25, 26]. One amazing result of these studies is the unbelievable sensitivity of insects to pheromones. Some deltoid moths attract mates with airborne pheromones from a distance of up to ten kilometres away.

Ants are similarly highly sensitive. Social insects possess a large arsenal of different pheromones allowing differentiated communication within the colony. The scent of some ant species is so effective that from the amount a single ant produces, one could lay a functional recruiting trail around the entire globe [118].

Pheromones in honeybees fulfil a number of functions [66, 204]. Butler, at the beginning of the seventeenth century, observed behavioural reactions of flying honeybees to scent originating from a distant source (see p. 10).

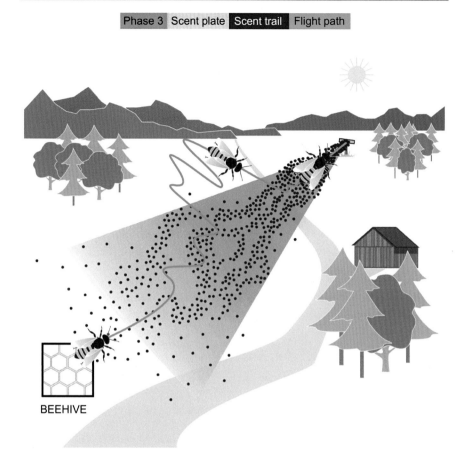

Fig. 12.1 The blue line shows the possible flight path of a bee that approaches a scent source. Dark blue points show the odour plume from a site. The plume structure is determined by air currents carrying scent molecules (after 10). If, as shown here, wind carries scent from the source to the hive, recruited bees follow this plume directly to the feeder. They step directly into Phase 3 (ATTRACT), and the first two phases (SEND and SEARCH) fall away

Scent plays a decisive role in Phase 3 of distant goal orientation of forager bees to a site advertised in a dance, and the scent will distract bees even from entrained flight paths. Forager bees, conditioned to a particular feeder, will, on their flight path to this site, be deflected by a certain amount if a familiar scent is brought to them from a different direction by the wind [169].

Two kinds of scent sources with fundamentally different characteristics come into consideration for foragers flying to a goal advertised in a dance: the site odour and from other bees on buzzing flights.

Site associated scents such as from flowers or artificially scented feeders release continuous odour plumes, and flow dynamics primarily determine their spatial and temporal distribution. In air, these are wind and convection currents for which exact measurements and models are available [10]. Flight paths of honeybees searching for a feeder in a wind tunnel show the area bees search corresponds well with where the scent spreads. This area is searched as soon as they happen to fly into the odour plume [124; see also Fig. 12.1].

Bees May Release Scent Waves into the Air with Their Buzzing Flights

Honeybees performing buzzing flights create a completely different kind of odour source. It is not associated with a fixed location and the odour does not spread like a scent plume coming from a flower.

Because buzzing flights only take place if bees had previously danced in the hive, and this is closely correlated with eversion of the Nasanov gland, it follows that the three behavioural patterns should be considered together.

Buzzing flights may cause the formation of odour "rings" in the air behind a flying bee, like those measured behind butterflies [85; see Fig. 12.2]. The formation of diffusing vortices near the boundary layer of fly wings is discussed by Werner Nachtigall [180 and personal communication] in relation to audible changes in flight tones of these insects. Such an audible alteration of flight tone led to the name "buzzing flight" in honeybees.

Such vortices are temporally and spatially stable, could trap the pheromone of the Nasanov gland, and result in a trail of scented rings in the air behind a flying bee. Measurements of similar events in water are available [17], showing such vortices in this dense medium to be stable for minutes. Seals follow their prey by swimming after the vortex trail fish leave behind them.

Do Optical Effects Serve Communication Between Bees?

In addition to orientation to an attractive scent, perhaps released by experienced bees in flight, it is possible experienced bees also provide a visual goal orientation aid.

Visual goal orientation is most likely for a swarm's flight to their final goal, in natural circumstances usually a hollow tree [14]. The rapid back and forth scout bee flights in the upper region of a swarm cloud [106] should be visible against the bright sky to most bees in the swarm that do not know the location of the goal. The obvious solution for swarm bees is to fly after the scouts to reach the goal.

Flying bees are not only perceivable as dark objects against the bright sky, but also their wings reflect light, including the ultraviolet component of sunlight. Bee wings trace a complex three-dimensional path during flight, beating at about 270

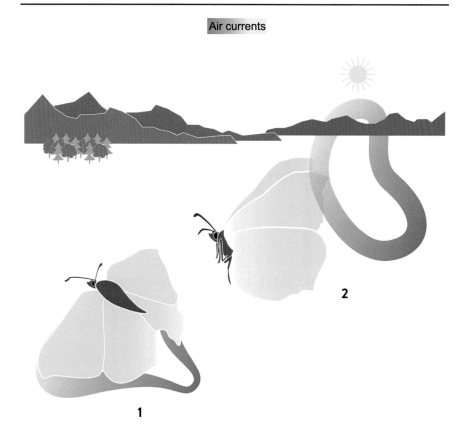

Fig. 12.2 The up and down motion of butterfly wings produce air currents in the form of rings that lie behind the flying insect [85]. Such rings in which the pheromone of the Nasanov gland is trapped may form during buzzing flights of honeybees resulting in an odour trail in the air. This, however, is not known

times a second, and bright reflections would be visible over a wide area. Three visual receptor cell types are responsible for colour vision in honeybees, each maximally sensitive to either green, blue, or ultraviolet (Fig. 12.3).

These thoughts about additional goal orientation aids for bees are speculative and mentioned here as an inducement for future research in this area.

Foragers land preferably on flowers where other bees are sitting [158] and perhaps provide a visual cue. Bees can visually distinguish quantities of up to four [108] and so in principle at least recognise small groups of bees.

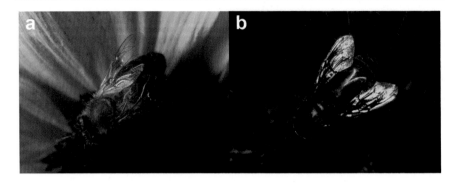

Fig. 12.3 Reflection of light from the wings of honeybees in daylight (**a**) and from short wavelength UV (**b**)

Studies with Polarised Light and Horizontal Hives

Honeybees orient their dances in the dark hive in relation to gravity. If they are simultaneously provided with a source of light, both stimuli affect the dance orientation (see Gould's experiments, p. 79ff). If a comb on which dances are held is laid down horizontally, the dances in the dark hive become completely disoriented because gravity falls away as an orientation reference. Waggle phases no longer indicate a specific angular region, but instead point randomly in all directions. In such a situation, bees will use a light source as the only possible orientation reference.

Although Gould used a point light source in his experiments, an alternative method employing polarised light offers a more precise and controllable optical stimulus. A small lamp in the hive appears at a different angle when seen from different positions on the comb (parallax) and is of limited use as an orientation reference.

The atmosphere of our earth acts like a polarising filter for sunlight and so the randomly spread oscillation planes of sunlight that reach the earth are "polarised" or ordered. Honeybees can detect this polarisation [60, 142, 214, 215, 278] and use it to orient their flights outside the hive. A view of a small section of the polarisation pattern of the sky is enough for them [278; first discovered by Karl von Frisch: 78] (Fig. 12.4).

The sensitivity of bees to polarised light was used in a series of experiments carried out on the dance language that Christine Scholtyssek submitted for her diploma thesis [221] to the University of Freiburg in 1998. The basis of this work was an elegant study thought up by Rüdiger Wehner and Samual Rossel for a research project on polarised light vision in honeybees [279].

The following experiment, we called the Gieshugel experiment, was carried out under similar conditions at a Beestation outpost of the University of Würzburg.

Fig. 12.4 The star foil, an original von Frisch device. Two overlapping discs, each consisting of eight segments of polarising foil arranged in a star rotate in opposite directions against one another. Turning the discs while looking through them at the sky reveals its polarisation pattern

The Gieshugel Experiment Provides Information About Behaviour of Bees in the Field

The Gieshugel experiment set out to examine how strongly the dance language influences recruit goal finding in the field. The basic idea of this study followed up on an experiment that Wehner and his co-workers had carried out years before [280]. Before publication in a professional journal, it was planned to repeat the study and increase the data range. Circumstances had previously prevented this and the results are now published here for the first time.

For the experiment, an observation hive was laid down horizontally and enclosed in a light tight hemisphere with a round aperture cut out of its highest point. A polarising foil covered the aperture (see Fig. 12.5). The foil allowed only the polarised portion of the skylight to fall on the comb providing bees with directional information. However, the information about the sun's position was ambiguous and no indication of which of the two possibilities was correct. One can compare the situation to parallel lines with arrowheads at both ends. For bees, it was not clear whether the sun stood on one side or exactly on the opposite side.

The bee's confusion about the true position of the sun led to the hoped-for and expected result. In every case, half of the waggle dances oriented to the polarisation pattern indicated one direction and the other half, the exact opposite direction (Fig. 12.6).

Marked foragers trained to feeder F1, advertised for feeder F1 but also for feeder F2, located exactly opposite at the same distance from the hive as F1 and to which none of the marked foragers had flown (for the experimental setup, see Fig. 12.7).

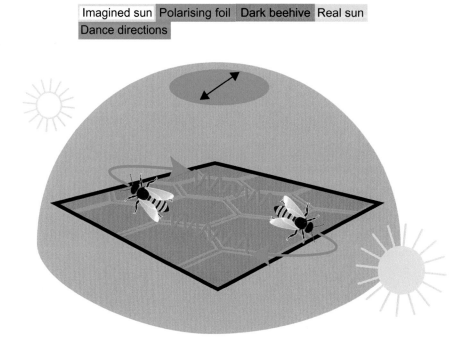

Imagined sun | Polarising foil | Dark beehive | Real sun
Dance directions

Fig. 12.5 The experimental setup of the Gieshugel experiment. The hive was laid horizontally on its side (so that orientation to gravity falls away) and enclosed by a light tight hemisphere with a round aperture cut out of its highest point and covered with a polarising foil (blue disc) to allow only polarised light to enter the hive. The light falling on the comb fitted two possible positions of the sun (double-headed arrow) but in opposite directions. The green waved lines show dances of bees which are in opposite directions because both are oriented to the polarised light

To ensure that both sites were equally attractive, foragers from a second, normally housed colony (with vertical combs) were trained to the advertised but did not visit site F2, for which they danced and recruited. In addition, a third feeder, F3, was set up, at the same distance away from the horizontally lying hive as F1 and F2. Bees from a second normal hive were also trained to F3.

This control station, not advertised in the experimental colony, was to gauge the proportion of foragers in the experimental colony that did not follow the instructions of dancers on the horizontal combs, nor the communication of the dancers outside the hive. They instead found the F3 feeder by randomly searching the area.

All three feeders offered sugar water and, except for one experiment (3 in Fig. 12.8), were not artificially scented. Recruits arriving at the three feeders were captured.

To determine the hive origin of recruits at feeders, *Apis mellifera* Hybrid var. Buckfast, a cross between several races, was employed as the experimental colony. The two control colonies were bees belonging to the race *Apis mellifera carnica*. The two races of bees are reliably distinguishable by their different colour.

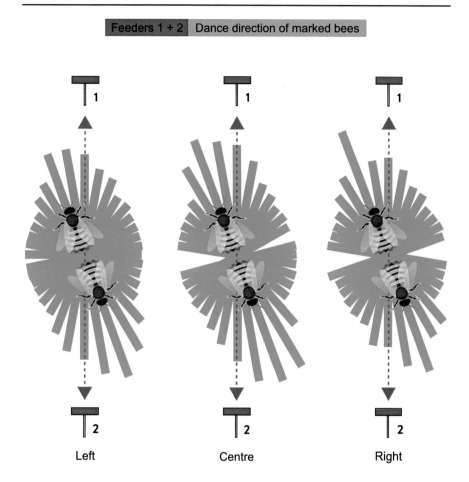

Fig. 12.6 Directions of waggle phases of all the dancing bees in the Gieshugel experiment. They are arranged in three categories, each after the position of the dancer in relation to the polarising foil (left, centre, right) because the area directly below the polarising foil was smaller than the area on which the dances were performed. The arrows show the exact direction of the feeders F1 and F2. The length of the green bars shows how often the waggle phases were oriented in the corresponding directions. In all, 5,255 waggle phases were counted. 2,624 (fifty per cent) left of the foil, 1,248 (23.7%) centred, and 1,383 (26.3%) right of the foil. The dances are clearly bimodal and bees indicate the two oppositely located feeders with the same frequency

The study produced "lying" (i.e. misled) bees because only one of the two directions shown was correct and indicated where the advertised feeder was actually located. The aim of the study was to examine how recruits distributed themselves at both of the sources indicated in the dance.

Feeders 1,2,3 Flight direction of trained bees Fields

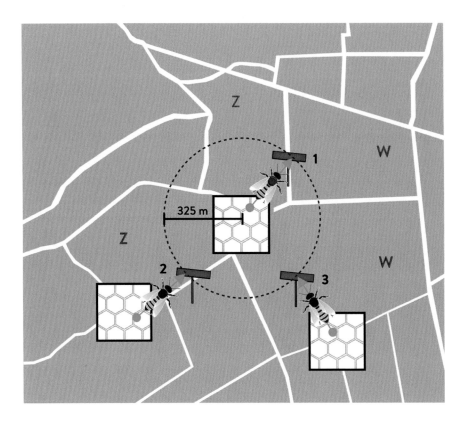

Fig. 12.7 Locations of the three hives and the three feeders in the Gieshugel experiment. Bases of the blue arrows show the exact position of the hives in the field and arrow tips show flight directions of marked bees from each hive to the respective feeders. All feeders were the same distance away from the hives (325 m: radius of the red circle). The surrounding fields carried crops of wheat (W) or sugar beet (Z)

What the Experiment Tells Us

An experiment of this nature allows predictions about different assumptions related to the mechanisms of recruitment. The dance language should distribute the recruits evenly to both indicated feeders. However, if a three-phase distant goal orientation takes place, the majority of the recruits should arrive at the feeder visited by the dancers.

	Feeder 1	Feeder 2	Feeder 3
Experiment 1	15	1	1
Experiment 2	40	6	1
Experiment 3	40	20	2

Fig. 12.8 The table shows the distribution of recruits to feeders in the Gieshugel experiment. Feeders 1 and 2 were located in exactly opposite directions and both advertised in dances although the experienced bees from the experimental colony flew only to feeder 1. Only bees from a different colony visited feeder 2 and feeder 3, set up as a control. Unmarked newcomers from the experimental colony that arrived at all three feeders during the experiment were counted and captured. Feeders were not scented in studies 1 and 2, and each feeder was scented with five drops of peppermint oil in study 3

The above table (Fig. 12.8) lists the distribution of recruits from the experimental colony to the three feeders. Each of these three studies was carried out on a separate day over a period of three hours.

The distribution of recruits to feeders F1 and F2 in all three studies did not reflect the equal division of the dance indication. In experiment three, where scent was employed, more recruits from the experimental colony now also arrived at the scented feeder 2, which although indicated in the dance was not visited by the experimental colony dancer. The dancers in experiment three carried peppermint scent that recruits searched for in the field during Phase 2 of the distant goal orientation. In all studies, most of the recruits arrived at F1, also visited by the dancer.

Wehner and his co-workers wrote the following about the method that produced the bimodal distribution of dances in the Gieshugel experiment:

"Here we show that bees really use the spatial information encoded in the direction and velocity of the waggle distances, fly to the location indicated by the dancing bee, but finally appear at the food source only when some additional requirements (presence of other bees, odour) are fulfilled…" [280, p. 1223].

This conclusion corresponds qualitatively with those of the Gieshugel experiments. A quantitative comparison of results from the two investigations is not possible. Wehner's publication [280] does not provide the numerical distribution of recruits at the two feeders indicated in the dances.

Pursuing a completely different question and experimental approach, Harald Esch also produced dances in which bees danced in two different directions. Esch trained a single dancer in a step study to a feeder that lay exactly in the direction of the sun. No information about the control stations is provided. In her dance, this bee made no return runs, instead a waggle phase followed directly on that before it. Esch achieved this in the following way: When the bee returned to dance in the hive and directed her waggle phase, as expected, straight up, Esch reflected an image of the sun from the bottom of the hive with a mirror. The dancer responded by following the first vertically upward waggle phase with one that headed vertically down, and without a return run. By continuously switching the mirror, up to 200 waggle phases followed one after the other. In the only study of this kind that Esch carried out with ambiguous dances [55], recruits flew exclusively to the feeder also visited by the dancer. The experiment has not been repeated but its observations confirm the Gieshugel experiments.

Conclusions from the Gieshugel experiment can be considered from the point of view of a three-phase distant goal orientation. Accordingly, recruits would, after following the ambiguous dances, fly to both the indicated areas where the search phase would begin. Recruits that set out in the correct direction would come across experienced bees flying back and forth. The data show that most recruits arrived at the feeder also visited by the dancers. Some recruits from the experimental colony landed where bees from another normal colony attracted their own recruits (see Fig. 12.8). One may speculate that these experimental colony recruits met the foreign bees during their search flights and joined them. Study 3 also showed that scent attracts significantly more recruits to the site indicated in the dance, but not visited by the dancer. Experimental conditions with windborne scents from various directions result in varying numbers of recruits being led to "wrong" but scented feeders.

What Can the Forest Habitat Contribute to the Research of Orientation and Communication of Honeybees?

Forests are a natural habitat for honeybees [6], in which natural selection ensured that the characteristics and abilities of bees enabled them to live and reproduce there. One look at a forest (Fig. 12.9) and it is very clear that the conditions here for orientation and communication of honeybees are completely different to those under which research into their orientation and communication have taken place (Fig. 12.10).

Essentially, three conditions in forests differ from those in an open field:

1. The forest environment is fractionated; there is no wide, distant view and no obvious landmarks for humans ("we cannot see the wood for the trees").
2. The view above is not of the free sky but instead appears as patches spread among the treetops.
3. A characteristic of the forest is, among others, that it is windless.

Abilities and qualities developed in the passage of evolution enable animals that live there to orient themselves and in the case of honeybees to communicate under these conditions.

So far, there has not been a single study of the orientation and communication of honeybees in their natural environment, the forest habitat. This is understandable given fundamental methodological problems making it unlikely that research in this area will occur with the same intensity as in open landscapes.

Fig. 12.9 A glimpse of a forest, a natural habitat of honeybees, makes it plain that not only honeybees have a problem of finding their way, which they have apparently solved, but also scarcely solvable tasks confront bee researchers

At this time, one can only speculate on the possible effects of unique conditions in forests on the orientation and communication of bees.

A distant view is excluded in a forest; one moves from place to place and to our eyes these places do not seem much different. Useful landmarks here are not obvious. Flights of bees over the tops of trees are possible, the sun serving as an orientation aid to find their home tree. Orientation using the polarisation pattern of the sky offers different conditions to that for flight beneath the open sky (Fig. 12.11).

Honeybees can direct long flights using the sun as a reference point. As Karl von Frisch discovered [78], in the absence of a direct view of the sun, bees use their ability to distinguish between polarised and unpolarised light to identify the polarisation pattern of the sky [279]. The extent and direction of polarisation changes evenly across the sky. Locations lying close together differ little in the polarisation extent and direction.

Bees flying through the forest instead of over the trees do not have a continuous sky above them, instead see a sky fragmented into small or larger patches (Fig. 12.12). The facets of the compound eyes of bees, directed towards the fragments of the sky during flight, can detect the intensity and direction of polarised light (see Fig. 5.2).

Fig. 12.10 Two experimental situations from step studies of Karl von Frisch. Two co-workers of von Frisch, each sitting in front of an observation hive. The flight path between the observation hive and the control stations, typical for countless experiments in the following decades of bee research, covered a stretch devoid of bushes and trees, with a wide and free view for the observers. Location and date of the photographs are not known

Fig. 12.11 Bees flying over the tops of trees should be in the position to find the tree in which their colony lives

Fig. 12.12 The section of compound eyes of honeybees directed towards the sky during flight detects the plane of polarised light. Under the open sky, the polarisation pattern changes continuously, in the forest it appears as broken patches. These patches, through their different polarisations, may help bees to orient and fly straight for long distances through even thickly wooded areas

Within the small patches of the sky, polarised light appears homogeneous. However, the polarisation is different in the different patches and increases with the distance between the patches. Bees would see a mosaic of patches with differently oriented polarisation planes.

What consequences this special optical panorama has for the orientation of bees in the forest is an open question. It could aid orientation and help them direct their flights over long distances even through thick forests. On the other hand, it could present difficulties and lend more weight to chemical orientation. Here there are questions upon questions.

The complexity of the communication behaviour of bees, their pronounced ability to learn, and the plasticity of their behaviour present particular challenges for the study of honeybees.

The integration of behavioural observations and experiments employing the most modern available methods will probably eventually provide a thorough and complete picture of the dance language of honeybees.

Correction to: The Struggle for Insight

Correction to:
Chapter 1 in: J. Tautz, *Communication Between Honeybees*,
https://doi.org/10.1007/978-3-030-99484-6_1

In the original version of the book, extra server material has been included and extra link has been provided in the copyright page. The erratum book has been updated with the changes.

The updated original version of this chapter can be found at
https://doi.org/10.1007/978-3-030-99484-6_1

Conclusion

From Aristotle to Maeterlinck

Karl von Frisch titled his epochal book *Tanzsprache und Orientierung der Biene* (*Dance language and orientation of bees*), thereby naming both the most impressive behavioural abilities of animals: Communication and spatial orientation.

Recruitment of foragers to a site is based on these two abilities. Bees do not differ in their orientation performance from many other insects. The possession of a dance language, however, apparently raises them to a level of their own, particularly if the classical concept of the function of the dance language is correct, namely that it indicates the position of a goal. "Only humans and honeybees can communicate factual information about distant localities"—such amazing conclusions are frequently found in relation to the dance language.

Exactly such an elevation of the honeybees may be the reason why every influence on orientation outside the hive that is suggested meets with resistance. This appears simple-minded, given, as expressed in this book, the absence of persuasive evidence that the dance sends a recruit directly to a goal.

The recruitment behaviour of bees is astonishing, and the meaning of observations extends beyond biological areas so it is no surprise the bee dance is one of best researched behaviours of animals. The occupation with this phenomenon over the last almost one hundred years has produced a veritable mountain of observations and data. The conclusions, though, do not exactly correspond with the facts. One comes across inconsistencies and contradictions that are resolved, however, if one starts out from the facts and not the interpretations. A logical view results from ordering the facts into the concept of distant goal orientation.

Near goal or direct orientation occurs if animals orienting to a goal reach it through signals or stimuli coming from the goal. In contrast, a distant goal, or indirect orientation, is primarily a chain of orientation events characterised by the lack of a direct connection with the goal, approached directly only over the last step of the way.

Near goal or direct orientation can be available to honeybees setting out from their hive for a flowering site, independent of its distance, if the wind carries the scent of flowers, even from far away, directly to the hive entrance. Bees can follow this message straight to the goal. This situation is mostly an exception. As a rule,

J. Tautz, *Communication Between Honeybees*, https://doi.org/10.1007/978-3-030-99484-6

workers flying out from the hive have no contact with the goal, and those that arrive at the site exhibit a distant goal or indirect orientation. For this, the bee dance (dance language) is indispensable for the first stage of goal discovery.

Karl von Frisch, as shown in his earlier studies of buzzing flights and the employment of Nasanov glands, was concerned with the last phase, Phase 3, of distant goal orientation and contributed important insights. Researchers who denied the function of the dance also contributed many details about Phase 3 of the distant goal orientation (above all, Adrian Wenner). Later works of Karl von Frisch and other scientists on the dance language, on the other hand, concerned themselves exclusively with Phase 1, the entry to the distant goal orientation sequence. The model that emerged from the studies made the dance language responsible for the entire path from hive to flowers, not merely the initial stage of a distant goal orientation. However, those that denied the importance of the dance and followed Wenner proceeded similarly blinkered and saw the orientation aids they researched to be effective over the entire way from start to finish. Both standpoints represented half the truth.

The von Frisch as well as the Wenner schools were, to a point, correct and incorrect in their concepts. Von Frisch was right that the dance exerts a strong influence on where dance followers direct their flights, but was wrong about the dance sending recruits directly to the goal. Wenner was right that scent alone, without dances, could bring recruits to the goal, but wrong about the dance offering no orientation aid for recruits.

So it was that one of the best-known controversies in behavioural biology came about, for which this book suggests a solution.

At the same time, by way of a new view of old data and their inclusion in a concept of distant goal orientation makes it clear how large the gaps in our knowledge are about the language of the bees.

Appendix

How to train bees to a feeder?

Honeybees have characteristics and abilities that make them ideal experimental animals. Karl von Frisch was the first to recognise and use it.

Bees possess excellent learning abilities helping them to distinguish between many different kinds of flowers by their colour, shape, and scent. Bees also have a good locational memory and a reliable sense of time.

In the field, bees quickly learn which flowers offer nectar at certain sites and times of day. Bees also quickly learn what is important at artificial feeders provided for them.

One can erect a feeder out in the field (see Fig. A.1), leave it, and wait until a searching forager finds it.

However, one can also control the process as follows: Containers are suitable as feeding stations if the presented honey or sugar water can be taken up as from flowers. Open liquid surfaces are less satisfactory than fine grooves in which the liquid can spread. The feeders are more conspicuous if coloured and scented, or contain scented food.

To train bees, a prepared feeder is set up directly in front of the hive so that foragers coming out of the hive do not have to fly but can discover the feeder by walking to it. The first bee to arrive at the feeder is marked with a spot of colour on its abdomen. Marked bees can be individually distinguished with different colours or patterns of spots. This makes it possible to identify bees returning in the course of time to the feeder from newcomers. Once marked bees return regularly to a feeder, one can begin to gradually move it away from the hive. Initially, the feeder should be moved only a short distance, but bees now have to fly to it. A waiting period for several rounds of visits from marked bees is required. When they return reliably, one can transport the feeder, with the bees sitting on it, a few metres further from the hive. Bees fly from the relocated feeder back to the hive after drinking, remember the position of the feeder, and return to it.

The above adequately describes the procedure. Feeders can be moved in increasingly larger steps on the way to the final location proposed for an experiment and it is possible to cover a distance of several hundred metres in half a day, although one should not proceed too rapidly. Obstructions along the path, such as bushes, can cause problems. Bees are distracted and difficult to keep in the study.

J. Tautz, *Communication Between Honeybees*, https://doi.org/10.1007/978-3-030-99484-6

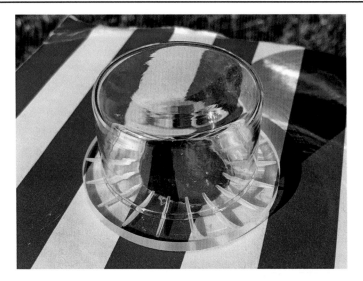

Fig. A.1 Forager bees take up sugar water flowing in small grooves in a plate beneath a glass container (4 cm in diameter). Coloured marking and artificial scenting of feeders belong to standard training experiments with honeybees

If it is intended to keep trained bees in a study for several days, food must be offered daily at the site they have learned. Should this not occur, bees quickly notice that this source no longer supplied food and do not return.

Glossary of Important Terms

Buzzing flights Forager bees that advertise for a food site with a waggle dance in the hive approach the indicated goal with a remarkable audible flight tone.

Comb-wide web The upper edges of comb cells form a net over which bees exchange vibration signals.

Compound eye A type of eye found in insects that is composed of many facets.

Cuticle The outer coating of the insect exoskeleton.

Dance followers Forager bees that follow the dance of a dancer in the hive.

Distant goal or indirect orientation An orientation process consisting of three phases in which the animal has no contact with the goal until the last phase.

Experienced bees Forager bees that are familiar with a food site.

Facets Single eyes of a compound eye.

Flagellum Also antennal ramus, the multi-segmented last part of the insect feeler.

Geraniol A mixture of alcohols and an important component of the Nasanov gland pheromone of bees.

Gravity sense organs Sense organs that detect the direction of gravity.

Johnstone organ A sense organ within the feelers of insects, highly sensitive for the detection of light air movements that stimulate the antennal flagellum.

Laser-Doppler vibrometry A modern technique allowing remote detection of the weak vibration of combs produced by dancers.

Mechanoreceptor Sensory receptor cells that detect mechanical displacement.

Nasanov gland A gland in the abdomen of worker bees producing a pheromone that attracts hive colleagues.

Near goal or direct orientation Orientation to a goal using signals emanating from the goal point.

J. Tautz, *Communication Between Honeybees*,
https://doi.org/10.1007/978-3-030-99484-6

Newcomers Forager bees that visit a food source for the first time.

Ocelli See facets.

Oscillation amplitude The largest excursion of an oscillation in relation to its resting point.

Pheromone Chemical compounds synthesised by living organisms and function as signals.

Polarisation The oscillation planes of light waves from the sun are randomly distributed. Light in which the oscillation plane is in only one direction is polarised. Light can be polarised with an optical filter.

Recruits/recruitment Recruits are forager bees that follow a dance in the hive and then arrive at the goal indicated by the dancer. Recruitment is the procedure for winning new foragers.

Saccade A sudden movement followed by a pause.

Sun azimuth-direction The point on the surface of the earth touched by an imaginary line drawn vertically down from the sun.

Tergite pocket glands Glands in each segment of a queen bee's abdomen that excrete queen pheromone.

Vibration pulse Short vibration bursts produced by the dancer during the dance.

Literature

1. Ai, H., Nishino, H., Itoh, T.: Topographic organization of sensory afferents of Johnston's organ in the honeybee brain. J. Comp. Neurol. **502**, 1030–1046 (2007)
2. Ai, H., Kai, K., Kumaraswamy, A., Ikeno, H., Wachtler, T.: Interneurons in the honeybee primary auditory center responding to waggle dance-like vibration pulses. J. Neurosci. **37**, 10624–10635 (2017)
3. Ai, H., Okada, R., Sakura, M., Wachtler, T., Ikeno, H.: Neuroethology of the waggle dance: how followers interact with the waggle dancer and detect spatial information. Insects **10**, 336 (2019)
4. Allison, J.D., Carde, R.T. (eds.): Pheromone Communication in Moths: Evolution, Behavior, and Application. University of California Press, Oakland (2016)
5. Aristoteles: Historia Animalium. Buch VIII und IX. Translated and explained by Stefan Schnieders. Walter de Gruyter, Berlin/Boston (2019)
6. Arndt, I., Tautz, J.: Honigbienen: Geheimnisvolle Waldbewohner. Knesebeck Verlag, München (2020)
7. Avitabile, A., Morse, R.A., Bloch, R.: Swarming honey bees guided by pheromones. Ann. Entomol. Soc. Am. **68**, 1079–1082 (1975)
8. Bacon, S.F.: Novum Organum (1629). In: Hutchins, R.M. (ed.) Great Books of Western World, pp. 103–195. Ill. Encyclopedia Britannica, Chicago (1952)
9. Batschelet, E.: Circular Statistics in Biology. Academic Press, London (1981)
10. Bau, J., Carde, R.T.: Modeling optimal strategies for finding a resource-linked, windborne odor plume: theories, robotics, and biomimetic lessons from flying insects. Integr. Comp. Biol. **55**, 461–477 (2015)
11. Beekman, M., Lew, J.B.: Foraging in honeybees—when does it pay to dance? Behav. Ecol. **19**, 255–261 (2008)
12. Beekman, M., Ratnieks, F.L.W.: Long-range foraging by the honey-bee, *Apis mellifera* L. Funct. Ecol. **14**, 490–496 (2000)
13. Beekman, M., Doyen, L., Oldroyd, B.P.: Increase in dance Imprecision with decreasing foraging distance in the honey-bee *Apis mellifera* L. is partly explained by physical constraints. J. Comp. Physiol. A **191**, 1107–1113 (2005)
14. Beekman, M., Fathke, R.L., Seeley, T.D.: How does an informed minority of scouts guide a honeybee swarm as it flies to its new home? Anim. Behav. **71**, 161–171 (2006)
15. Beekman, M., Sumpter, D.J.T., Seraphides, N., Ratnieks, F.L.W.: Comparing foraging behaviour of small and large honey-bee colonies by decoding waggle dances made by foragers. Funct. Ecol. **18**, 829–835 (2004)
16. Biesmeijer, J.C., Seeley, T.D.: The use of waggle dance information by honeybee throughout their foraging careers. Behav. Ecol. Sociobiol. **59**, 133–142 (2005)
17. Bleckmann, H., Breithaupt, T., Blickhan, R., Tautz, J.: The time course and frequency content of hydrodynamic events caused by moving fish, frogs, and crustaceans. J. Comp. Physiol. A **168**, 749–757 (1991)

© The Editor(s) (if applicable) and The Author(s), under exclusive license to Springer   153
Nature Switzerland AG 2022
J. Tautz, *Communication Between Honeybees*,
https://doi.org/10.1007/978-3-030-99484-6

18. Boch, R.: Die Tänze der Bienen bei nahen und fernen Trachtquellen. Z. vergl. Physiol. **38**, 136–167 (1956)
19. Boch, R.: Rassenmässige Unterschiede bei den Tänzen der Honigbiene (*Apis mellifica* L.). Z. vergl. Physiol. **40**, 289–320 (1957)
20. Bonnier, G.: Sur la division du travail chez les abeilles. CR Acad. Sic. Paris **143**, 941–946 (1906)
21. Božič, J., Valentinčič, T.: Attendants and followers of honey bee waggle dances. J. Apic. Res. **30**, 125–131 (1991)
22. Breed, M.D.: Recognition pheromones of the honey bee. Bioscience **48**, 463–470 (1998)
23. Brockmann, A., Robinson, G.E.: Central projections of sensory systems involved in honey bee dance language communication. Brain Behav. Evol. **70**, 125–136 (2007)
24. Butler, C.: The Feminine Monarchie or a Treatise Concerning Bees. Joseph Barnes, Oxford (1609)
25. Carde, R.T.: Chemo-orientation in flying insects. In: Bell, D.J., Carde, R.T. (eds.) Chemical Ecology of Insects, pp. 111–124. Chapman & Hall, New York (1984)
26. Carde, R.T.: Moth navigation along pheromone plumes. In: Allison, J.D., Carde, R.T. (eds.) Pheromone Communication in Moths: Evolution, Behavior, and Application, pp. 173–189. University of California Press, Oakland (2016)
27. Chatterjee, A., George, E.A., Prabhudev, M.V., Basu, P., Brockmann, A.: Honey bees flexibly use two navigational memories when updating dance distance information. J. Exp. Biol. **222**, 1–11 (2019)
28. Cheeseman, J.F., Millar, C.D., Greggers, U., Lehmann, K., Pawley, M.D.M., Gallistel, C.R., Warman, G.R., Menzel, R.: Way-finding in displaced clock-shifted bees proves bees use a cognitive map. PNAS USA **111**, 8949–8954 (2014)
29. Cheeseman, J.F., Millar, C.D., Greggers, U., Lehmann, K., Pawley, M.D.M., Gallistel, C.R., Warman, G.R., Menzel, R.: Reply to Cheung et al.: the cognitive map hypothesis remains the best Interpretation of the data in honeybee navigation. PNAS USA **111**, E4398 (2014)
30. Cheung, A., Collett, M., Collett, T.S., Dewar, A., Dyer, F., Graham, P., Mangan, M., Narendra, A., Philippides, A., Sturzl, W., Webb, B., Wystrach, A., Zeil, J.: Still no convincing evidence for cognitive map use by honeybees. PNAS USA **111**, E4396–E4397 (2014)
31. Christ, J.L.: Anweisung zur nützlichsten und angenehmsten Bienenzucht. Johann Benjamin Georg Fleischer, Leipzig (1783)
32. Cochran, W.W., Wikelski, M.: Individual migratory tactics of New World *Catharus* thrushes: current knowledge and future tracking Options from space. In: Greenberg, R., Marra, P.P. (eds.) Birds of Two Worlds, pp. 274–289. John Hopkins University Press, Baltimore (2005)
33. Couvillon, M.J.: The dance legacy of Karl von Frisch. Insect. Soc. **59**, 297–306 (2012)
34. Couvillon, M.J., Pearce, F.C.R., Harris-Jones, E.L., Kuepfer, A.M., Mackenzie-Smith, S.J., Rozario, L.A., Schurch, R., Ratnieks, F.L.W.: Intra-dance variation among waggle runs and the design of efficient protocols for honey bee dance decoding. Biol. Open **15**, 467–472 (2012)
35. Couvillon, M.J., Schurch, R., Ratnieks, F.L.W.: Dancing bees communicate a foraging preference for rural lands in high level agri-environment schemes. Curr. Biol. **24**, 11, 1212–1215 (2014)
36. Couvillon, M.J., Schurch, R., Ratnieks, F.L.W.: Waggle dance distances as integrative indicators of seasonal foraging challenges. PLOS ONE **9**, e93495 (2014)
37. Cruse, H.J., Wehner, R.: No need for a cognitive map: decentralized memory for insect navigation. PLoS. Comput. Biol. **7**, e1002009 (2011)
38. Dawkins, R.: Bees are easily distracted. Science **165**, 751 (1969)
39. De Marco, R.J., Farina, W.M.: Changes in food source profitability affect the trophallactic and dance behavior of forager honeybees (*Apis mellifera* L.). Behav. Ecol. Sociobiol. **50**, 441–449 (2001)

40. De Marco, R.J., Menzel, R.: Learning and memory in communication and navigation in insects. In: Menzel, R., Byrne, J. (eds.) Learning and Memory: A Comprehensive Reference, pp. 477–498. Elsevier, New York (2008)

41. De Marco, R.J., Gil, M., Farina, W.M.: Does an increase in reward affect the precision of the encoding of directional information in the honeybee waggle dance? J. Comp. Physiol. A **191**, 413–419 (2005)

42. De Marco, R.J., Gurevitz, J.M., Menzel, R.: Variability in the encoding of spatial information by dancing bees. J. Exp. Biol. **211**, 1635–1644 (2008)

43. Donaldson-Matasci, M.C., Dornhaus, A.: How habitat affects the benefits of communication in collectively foraging honey bees. Behav. Ecol. Sociobiol. **66**, 583–592 (2012)

44. Donaldson-Matasci, M.C., DeGrandi-Hoffman, G., Dornhaus, A.: Bigger is better: honeybee colonies as distributed information-gathering systems. Anim. Behav. **85**, 585–592 (2013)

45. Dornhaus, A., Chittka, L.: Why do honey bees dance? Behav. Ecol. Sociobiol. **55**, 395–401 (2004)

46. Dreller, C., Kirchner, W.H.: Hearing in honeybees: localization of the auditory sense organ. J. Comp. Physiol. A **173**, 275–279 (1993)

47. Dyer, F.C.: Memory and sun compensation by honey bees. J. Comp. Physiol. A **160**, 621–633 (1987)

48. Dyer, F.C.: Bees acquire route-based memories but not cognitive maps in a familiar landscape. Anim. Behav. **41**, 239–246 (1991)

49. Dyer, F.C.: The biology of the dance language. Annu. Rev. Entomol. **47**, 917–949 (2002)

50. Dyer, F.C.: Dance language. In: Breed, M.D. (ed.) Encyclopedia of Animal Behavior, pp. 445–453. Elsevier Acad. Press, Oxford (2010)

51. Esch, H.: Die Elemente der Entfernungsmitteilung im Tanz der Bienen. Experientia **12**, 439–441 (1956)

52. Esch, H.: Analyse der Schwänzelphase im Tanz der Bienen. Naturwissenschaften **9**, 207 (1956)

53. Esch, H.: Über die Schallerzeugung beim Werbetanz der Honigbiene. Z. vergl. Physiol. **45**, 1–11 (1961)

54. Esch, H.: Auswirkung der Futterplatzqualität auf die Schallerzeugung im Werbetanz der Honigbiene. Verh. Dtsch. Zool. Ges. **26**, 302–309 (1963)

55. Esch, H.: Beiträge zum Problem der Entfernungsweisung in den Schwänzeltänzen der Honigbienen. Z. vergl. Physiol. **48**, 534–546 (1964)

56. Esch, H., Bastian, J.A.: How do newly recruited honey bees approach a food site? Z. vergl. Physiol. **68**, 175–181 (1970)

57. Esch, H., Burns, J.: Distance estimation by foraging honeybees. J. Exp. Biol. **199**, 155–162 (1996)

58. Esch, H., Wilson, D.: The sounds produced by flies and bees. Z. vergl. Physiol. **54**, 256–267 (1967)

59. Esch, H., Zhang, S., Srinivasan, M., Tautz, J.: Honeybee dances communicate distance by optic flow. Nature **411**, 581–583 (2001)

60. Evangelista, C., Kraft, P., Dacke, M., Labhart, T., Srinivasan, M.V.: Honeybee navigation: critically examining the role of the polarization compass. Philos. Trans. R. Soc. B Biol. Sci. **369** (2014)

61. Farina, W.M.: Food-exchange by foragers in the hive—a means of communication among honey bees? Behav. Ecol. Sociobiol. **38**, 59–64 (1996)

62. Farina, W.M., Gruter, C., Diaz, P.C.: Social learning of floral odours inside the honeybee hive. Proc. R. Soc. B **272**, 1923–1928 (2005)

63. Fernandez, P.C., Gil, M., Farina, W.M.: Reward rate and forager activation in honeybees: recruiting mechanisms and temporal distribution of arrivals. Behav. Ecol. Sociobiol. **54**, 80–87 (2003)

64. Ferrari, T.E., Tautz, J.: Severe honey bee (*Apis mellifera*) losses correlate with geomagnetic disturbances in earth's atmosphere. J. Astrobiol. Outreach **134** (2015)

65. Francon, J.: Die Klugheit der Bienen. Paul Neff Verlag, Berlin (1939)

66. Free, J.B.: The conditions under which foraging honeybees expose their Nasanov glad. J. Apic. Res. **7**, 139–145 (1968)

67. Free, J.B., Williams, I.H.: Exposure of the Nasanov gland by honeybees (*Apis mellifera*) collecting water. Behaviour **37**, 286–290 (1970)

68. Friesen, L.J.: The search dynamics of recruited honeybees, *Apis mellifera ligustica* Spinola. Biol. Bull. **144**, 107–131 (1973)

69. Frisch, K.v.: Der Farbensinn und Formensinn der Bienen. Zool. Jb. Abt. Allg. Zool. u. Physiol. **35**, 1–188 (1914)

70. Frisch, K.v.: Über den Geruchsinn der Bienen und seine blütenbiologische Bedeutung. Zool. Jb. Abt. allg. Zool. u. Physiol. **37**, 1–238 (1919)

71. Frisch, K.v.: Über das Tanzen der Honigbiene. Bayerische Bienen-Zeitung **42**, 167–168 (1920)

72. Frisch, K.v.: Über die ≫Sprache≪ der Bienen. Zool. Jb. Abt. allg. Zool. u. Physiol. **40**, 1–186 (1923)

73. Frisch, K.v.: Sinnesphysiologie und ≫Sprache≪ der Bienen. Sonderausgabe der Naturwissenschaften **12**, 1–27 (1924)

74. Frisch, K.v.: Die Werbetänze der Bienen und ihre Auslösung. Naturwissenschaften **30**, 269–277 (1942)

75. Frisch, K.v.: Die Tänze der Bienen. Osterr. Zool. Z. **1**, 1–48 (1946)

76. Frisch, K.v.: Die ≫Sprache≪ der Bienen und ihre Nutzanwendung in der Landwirtschaft. Experientia **2**, 397–404 (1946)

77. Frisch, K.v.: Duftgelenkte Bienen im Dienste der Landwirtschaft und Imkerei. Salzburger Wochenschau für Österreich 25/26 (1946)

78. Frisch, K.v.: Tanzsprache und Orientierung der Bienen. Springer, Heidelberg (1965)

79. Frisch, K.v.: Honeybees: do they use direction and distance information provided by their dances? Science **158**, 1072–1076 (1967)

80. Frisch, K.v., Jander, R.: Über den Schwänzeltanz der Bienen. Z. vergl. Physiol. **40**, 239–263 (1957)

81. Frisch, K.v., Kratky, O.: Über die Beziehung zwischen Flugweite und Tanztempo bei der Entfernungsmeldung der Bienen. Naturwissenschaften **49**, 409–417 (1962)

82. Frisch, K.v., Lindauer, M.: The ≫language≪ and orientation of the honey bee. Annu. Rev. Entomol. **1**, 45–58 (1956)

83. Frisch, K.v., Lindauer, M.: Über die ≫Misweisung≪ bei den richtungsweisenden Tänzen der Bienen. Naturwissenschaften **18**, 585–594 (1961)

84. Frisch, K.v., Rösch, G.A.: Neue Versuche über die Bedeutung von Duftorgan und Pollenduft für die Verständigung im Bienenvolk. Z. vergl. Physiol. **4**, 1–21 (1926)

85. Fuchiwaki, M., Kuroki, T., Tanaka, K., Tababa, T.: Dynamic behaviour of the vortex ring formed on a butterfly wing. Exp. Fluids **54**, Article number: 1450 (2013)

86. Gagliardo, A., Ioale, P., Savini, M., Lipp, H.-P., Dell'Omo, D.: Finding home: the final step of the pigeons' homing process studied with a GPS data logger. J. Exp. Biol. **210**, 1132–1138 (2007)

87. Gaius Plinius Secundus: Naturalis historia (also Historia naturalis), Book XI: The various kinds of insects, Kapitel 16

88. Galizia, C.G., Sachse, S., Rappert, A., Menzel, R.: The glomerular code for odor representation is species specific in the honeybee *Apis mellifera*. Nat. Neurosci. **2**, 473–478 (1999)

89. Gardner, K.E., Seeley, T.D., Calderone, N.W.: Hypotheses on the adaptiveness or non-adaptiveness of the directional imprecision in the honey bee's waggle dance (Hymenoptera: Apidae: *Apis mellifera*). Entomol. Gen. **29**, 285–298 (2007)

90. Gardner, K.E., Seeley, T.D., Calderone, N.: Do honeybees have two discrete dances to advertise food sources? Anim. Behav. **75**, 1291–1300 (2008)

91. Gil, M., De Marco, R.J.: Decoding information in the honeybee dance: revisiting the tactile hypothesis. Anim. Behav. **80**, 887–894 (2010)
92. Gil, M., Farina, W.M.: Foraging reactivation in the honeybee *Apis mellifera* L.: factors affecting the return to known nectar sources. Naturwissenschaften **89**, 322–325 (2002)
93. Gilley, D.C.: Hydrocarbons emitted by waggle-dancing honey bees increase forager recruitment by stimulating dancing. PLOS ONE **9**(8): e105671 (2014)
94. Giurfa, M., Nunez, J.A.: Honeybees mark with scent and reject recently visited flowers. Oecologia **89**, 113–117 (1992)
95. Gould, J.L.: Honeybee communication: the dance-language controversy. Dissertation, Rockefeller University, New York (1975)
96. Gould, J.L.: Honey bee recruitment: the dance-language controversy. Science **189**, 685–693 (1975)
97. Gould, J.L.: The dance language controversy. Q. Rev. Biol. **51**, 211–244 (1976)
98. Gould, J.L.: Teaching bees to lie. Mosaic **10**, 31–32 (1979)
99. Gould, J.L.: The case for magnetic sensitivity in birds and bees (such as it is). Am. Sci. **68**, 256–267 (1980)
100. Gould, J.L.: Natural history of honey bee learning. In: Marler, P., Terrace, H.S. (eds.) The Biology of Learning, pp. 149–180. Springer, Berlin (1984)
101. Gould, J.L., Gould, C.G.: The insect mind: physics or metaphysics? In: D.R. Griffin (ed.) Animal Mind Human Mind, pp. 269–298. Springer, Heidelberg (1982)
102. Gould, J.L., Gould, C.G.: The Honey Bee. W. H. Freeman, New York (1988)
103. Gould, J.L., Towne, W.F.: Evolution of the dance language. Am. Nat. **130**, 317–338 (1987)
104. Gould, J.L., Dyer, F.C., Towne, W.F.: Recent progress in the study of the dance language. In: Hölldobler, B., Lindauer, M. (eds.) Fortschritte der Zoologie, 3: Experimental Behavioral Ecology, pp. 141–161. Fischer Verlag, Stuttgart (1985)
105. Gould, J.L., Henery, M., McLeod, M.C.: Communication of direction by the honey bees. Science **169**, 544–554 (1970)
106. Greggers, U., Schoning, C., Degen, J., Menzel, R.: Scouts behave as streakers in honeybee swarms. Naturwissenschaften **100**, 805–809 (2013)
107. Greggers, U., Koch, G., Schmidt, V., Durr, A., Floriou-Servou, A., Piepenbrock, D., Gopfert, M.C., Menzel, R.: Reception and learning of electric fields in bees. Proc. R. Soc. B **280**, 20130528 (2013)
108. Gross, H.J., Pahl, M., Si, A., Zhu, H., Tautz, J., Zhang, S.: Number-based visual generalisation in the honeybee. PLOS ONE **4**, e4263 (2009)
109. Grüter, C., Farina, W.M.: The honeybee waggle dance: can we follow the steps? Trends Ecol. Evol. **24**, 242–247 (2009)
110. Grüter, C., Leadbeater, E.: Insights from insects about adaptive social information use. Trends Ecol. Evol. **29**, 177–184 (2014)
111. Grüter, C., Ratnieks, F.L.W.: Honeybee foragers increase the use of waggle dance information when private information becomes unrewarding. Anim. Behav. **81**, 949–954 (2011)
112. Grüter, C., Sol Balbuena, M., Farina, W.M.: Informational conflicts created by the waggle dance. Proc. R. Soc. B. **275**, 1321–1327 (2008)
113. Haldane, J.B.S., Spurway, H.: A statistical analysis of communication in ≫*Apis mellifera*≪ and a comparison with communication in other animals. Insect. Soc. **1**, 247–283 (1954)
114. Harano, K., Mitsuhata-Asai, A., Konishi, T., Suzuki, T., Sasaki, M.: Honeybee foragers adjust crop contents before leaving the hive. Behav. Ecol. Sociobiol. **67**, 1169–1178 (2013)
115. Hasegawa, Y., Ikeno, H.: How do honeybees attract nestmates using waggle dances in dark and noisy hives? PLoS ONE **6**, e19619 (2011)
116. Hasenjager, M.J., Hoppitt, W., Leadbeater, E.: Network-based diffusion analysis reveals context-specific dominance of dance communication in foraging honeybees. Nat. Commun. **11**, Article number 625 (2020)
117. Heidborn, T.: Der Forscher, der auf Bienen flog. Max Planck Forschung **1**(10), 76–82 (2010)

118. Hölldobler, B., Wilson, E.O.: The Ants. Springer, Heidelberg (1990)

119. Hrncir, M., Barth, F.G., Tautz, J.: Acoustic communication in bees. In: Drosopoulos, S., Claridge, M.F. (eds.) Insect sounds and Communication: Physiology, Behaviour, Ecology, and Evolution. CRC-Press, Boca Raton (2005)

120. Hrncir, M., Gravel, A.I., Schorkopf, D.L.P., Schmidt, V.M., Zucchi, R., Barth, F.G.: Thoracic vibrations in stingless bees (*Melipona seminigra*): resonances of the thorax influence vibrations associated with flight but not those associated with sound production. J. Exp. Biol. **211**, 678–685 (2008)

121. Hrncir, M., Maia-Silva, C., Mc Cabe, S.I., Farina, W.M.: The recruiter's excitement— features of thoracic vibrations during the honey bee's waggle dance related to food source profitability. J. Exp. Biol. **214**, 4055–4064 (2011)

122. Huber, F.: Nouvelles observations sur les abeilles. Debray, Paris (1796). Neue Beobachtungen an den Bienen. Deutsch von G. Kleine. Ehlers, Einbeck (1856)

123. Hunt, J.H., Richard, F.J.: Intracolony vibroacoustic communication in social insects. Insect. Soc. **60**, 403–417 (2013)

124. Ikeno, H., Akamatsu, T., Hasegawa, Y., Ai, H.: Effect of olfactory stimulus on the flight course of a honeybee, *Apis mellifera*, in a wind tunnel. Insects **5**, 92–104 (2014)

125. Ikeno, H., Kumaraswamy, A., Kai, K., Wachtler, T., Ai, H.: A segmentation scheme for complex neuronal arbors and application to vibration sensitive neurons in the honeybee brain. Front. Neuroinform. 12/61 (2018)

126. Jammalamadaka, S.R., Sengupta, A.: Topics in Circular Statistics. Series on Multivariate Analysis, vol. 5. World Scientific, Singapore (2001)

127. Jensen, I.L., Michelsen, A., Lindauer, M.: On the directional indications in the round dances of honeybees. Naturwissenschaften **84**, 452–454 (1997)

128. Johannes von Salisbury: Metalogicon 3, 4, 47–50 (1159)

129. Johnson, D.L.: Communication among bees with field experience. Anim. Behav. **15**, 487–492 (1967)

130. Johnson, D.L.: Honey bees: do they use the direction information contained in their dance maneuver? Science **155**, 844–847 (1967)

131. Judd, T.M.: The waggle dance of the honey bee: which bees following a dancer successfully acquire the information? J. Insect Behav. **8**, 343–354 (1994)

132. Kietzman, P., Visscher, K.: Follower position does not affect waggle dance information transfer. Psyche: J. Entomol. 1–5 (2019)

133. Kilpinen, O., Storm, J.: Biophysics of the subgenual organ of the honeybee, *Apis mellifera*. J. Comp. Physiol. A **181**, 30–318 (1997)

134. Kirchner, W.H.: Hearing in honeybees: the mechanical response of the bee's antenna to near field sound. J. Comp. Physiol. A **175**, 261–265 (1994)

135. Kirchner, W.H., Grasser, A.: The significance of odor cues and dance language information for the food search behavior of honeybees (Hymenoptera: Apidae). J. Insect Behav. **11**, 169–178 (1998)

136. Kirchner, W.H., Towne, W.F.: The sensory basis of the honeybee's dance language. Sci. Am. **270**, 52–59 (1994)

137. Kirchner, W.H., Dreller, C., Towne, W.F.: Hearing in honeybees: operant conditioning and spontaneous reactions to airborne sound. J. Comp. Physiol. A **168**, 85–89 (1991)

138. Kirchner, W.H., Lindauer, M., Michelsen, A.: Honeybee dance communication: acoustical indication of direction in round dances. Naturwissenschaften **75**, 629–630 (1988)

139. Klir, G.J., Folger, T.A.: Fuzzy Sets, Uncertainty and Information. Prentice-Hall, New York (1988)

140. Knaffl, H.: Über die Flugweite und Entfernungsmeldung der Bienen. Z. Bienenforsch. **2**, 131–140 (1953)

141. Kohl, P.L., Thulasi, N., Rutschmann, B., George, E.A., Steffan-Dewenter, I., Brockmann, A.: Adaptive evolution of honeybee dance dialects. Proc. R. Soc. B **287**, 20200190 (2020)

142. Kraft, P., Evangelista, C., Dacke, M., Labhart, T., Srinivasan, M.V.: Honeybee navigation: following routes using polarized-light cues. Philos. Trans. R. Soc. B Biol. Sci. **366**, 703–708 (2011)

143. Kumaraswamy, A., Ai, H., Kai, K., Ikeno, H., Wachtler, T.: Adaptations during maturation in an identified honeybee interneuron responsive to waggle dance vibration signals. eNeuro **6** (2019)

144. Landgraf, T.: Robobee: a biomimetic honeybee robot for the analysis of the dance communication system. Dissertation, Freie Universität Berlin (2013)

145. Landgraf, T., Rojas, R., Nguyen, H.F., Kriegel, H.E., Stettin, K.: Analysis of the waggle dance motion of honeybees for the design of a biomimetic honeybee robot. PLOS ONE **6**, e21354 (2011)

146. Landgraf, T., Bierbach, D., Kirbach, A., Cusing, R., Oertel, M., Lehmann, K., Greggers, U., Menzel, R., Rojas, R.: Dancing honey bee robot elicits dance-following and recruits foragers (2018). arxiv.org/pdf/1803.07126v1

147. L'Anson Price, R., Grüter, C.: Why, when and where did honey bee dance communication evolve? Front. Ecol. Evol. **3**, 125 (2015)

148. L'Anson Price, R., Dulex, N., Vial, N., Vincent, C., Grüter, C.: Honeybees forage more successfully without the ≫danced language≪ in challenging environments. Sci. Adv. **5**, 1–9 (2019)

149. Lehrer, M. (ed.): Orientation and Communication in Arthropods. Springer, Heidelberg (1997)

150. Lindauer, M.: Schwarmbienen auf Wohnungssuche. Z. vergl. Physiol. **37**, 263–324 (1955)

151. Lindauer, M.: Orientierung im Erdmagnetfeld. Fortschr. Zool. **21**, 211–228 (1973)

152. Lindauer, M., Martin, H.: Magnetic effect on dancing bees. In: Galler, S.R., Schmidt-Koenig, K. (eds.) Animal Orientation and Navigation, pp. 559–567. Scient. and Techn. Information Off., Washington (1972)

153. Lindauer, M., Schricker, B.: Über die Funktion der Ocellen bei den Dämmerungsflügen der Honigbiene. Biol. Zbl. **82**, 721–725 (1963)

154. Linnaeus, C.V.: Horologium Florae. In: Philosophia Botanica, pp. 274–276. R. Kiesewetter, Stockholm (1751)

155. Lineburg, B.: Communication by scent in the honey-bee—a theory. Am. Nat. **58**, 530–537 (1924)

156. Łopuch, S., Tofilski, A.: Importance of wing movements for information transfer during honey bee waggle dance. Ethology **123**, 974–980 (2017)

157. Łopuch, S., Tofilski, A.: Impact of the quality of food sources on the wing beating of honey bee dancers. Apidologie **51**, 631–641 (2020)

158. Lowell, E.S.H., Morris, J.A., Vidal, M.C., Durso, C.S., Murphy, S.M.: The effect of conspecific cues on honey bee foraging behavior. Apidologie **50**, 454–462 (2019)

159. Lubbock, J.: Ants Bees, and Wasps. A Record of Observations on the Habits of the Social Hymenoptera. D. Appleton and Company, New York (1882)

160. Maeterlinck, M.: Das Leben der Bienen. Eugen Diederichs, Jena (1919)

161. Markl, H.: Ein neuer Propriorezeptor am Coxa-Trochanter-Gelenk der Honigbiene. Naturwissenschaften **52**, 460 (1965)

162. Markl, H.: Schwerkraftdressuren an Honigbienen. I. Die geomenotaktische Fehlorientierung. Z. vergl. Physiol. **53**, 328–352 (1966)

163. Markl, H.: Schwerkraftdressuren an Honigbienen. II. Die Rolle der schwererezeptorischen Borstenfelder verschiedener Gelenke für die Schwerekompassorientierung. Z. vergl. Physiol. **53**, 353–371 (1966)

164. Markl, H.: Manipulation, modulation, information, cognition: some of the riddles of communication. Fortschr. Zool. **31**, 163–194 (1985)

165. Martin, H., Lindauer, M.: Der Einfluss des Erdmagnetfeldes auf die Schwereorientierung der Honigbiene (*Apis mellifica*). J. Comp. Physiol. **122**, 145–187 (1977)

166. Mautz, D.: Der Kommunikationseffekt der Schwänzeltänze bei *Apis mellifica carnica* (Pollm.). Z. vergl. Physiol. **72**, 197–220 (1971)
167. Maynard Smith, J., Harper, D.: Animal Signals. Oxford University Press, Oxford (2003)
168. Menzel, R.: Unbeirrt zum Ziel. Deutsches Bienen-Journal **1**, 56–59 (2019)
169. Menzel, R., Greggers, U.: Guidance by odors in honeybee navigation. J. Comp. Physiol. A **199**, 867–873 (2013)
170. Menzel, R., Greggers, U., Smith, A., Berger, S., Brandt, R., Brunke, S., Bundrock, G., Huelse, S., Pluempe, T., Schaupp, F., Schuettler, E., Stach, S., Stindt, J., Stollhoff, N., Watzl, S.: Honeybees navigate according to a map-like spatial memory. PNAS USA **102**, 3040–3045 (2005)
171. Menzel, R., Kirbach, A., Haass, W.-D., Fischer, B., Fuchs, J., Koblofsky, M., Lehmann, K., Reiter, L., Meyer, H., Nguyen, H., Jones, S., Norton, P., Greggers, U.: A common frame of reference for learned and communicated vectors in honeybee navigation. Curr. Biol. **21**, 645–650 (2011)
172. Michelsen, A.: Ein mechanisches Modell der tanzenden Honigbiene. Mit einer kleinen Roboterbiene kann man die Sprache der Bienen studieren und den Bienen erzählen, wohin sie fliegen sollen. BiuZ **19**, 121–126 (1989)
173. Michelsen, A.: Signals and flexibility in the dance communication of honeybees. J. Comp. Physiol. A **189**, 165–174 (2003)
174. Michelsen, A., Andersen, B.B., Kirchner, W.H., Lindauer, M.: Honeybees can be recruited by a mechanical model of a dancing bee. Naturwissenschaften **76**, 277–280 (1989)
175. Michelsen, A., Andersen, B.B., Kirchner, W.H., Lindauer, M.: How honeybees perceive communication dances, studied by means of a mechanical model. Behav. Ecol. Sociobiol. **30**, 143–150 (1992)
176. Michelsen, A., Kirchner, W.H., Lindauer, M.: Sound and vibrational signals in the dance language of the honeybee, *Apis mellifera*. Behav. Ecol. Sociobiol. **18**, 207–212 (1986)
177. Michelsen, A., Towne, W.F., Kirchner, W.H., Kryger, P.: The acoustic near field of a dancing honeybee. J. Comp. Physiol. A **161**, 633–643 (1987)
178. Munz, T.: The bee battles: Karl von Frisch, Adrian Wenner and the honey-bee dance language controversy. J. Hist. Biol. **38**, 535–570 (2005)
179. Munz, T.: Der Tanz der Bienen: Karl von Frisch und die Entdeckung der Bienensprache. Czernin, Wien (2018)
180. Nachtigall, W.: Hohe Fluggeschwindigkeiten subalpiner Zweiflügler (Diptera). Entomol. Gen. **26**, 235–239 (2003)
181. Nieh, J.C., Tautz, J.: Behaviour-locked signal analysis reveals weak 200–300 Hz comb vibrations during the honeybee waggle dance. J. Exp. Biol. **203**, 1573–1579 (2000)
182. Nunez, J.A.: Quantitative Beziehungen zwischen den Eigenschaften von Futterquellen und dem Verhalten von Sammelbienen. Z. vergl. Physiol. **53**, 142–164 (1966)
183. Nunez, J.A.: Sammelbienen markieren versiegte Futterquellen durch Duft. Naturwissenschaften **54**, 322–323 (1967)
184. Nunez, J.A.: The relationship between sugar flow and foraging and recruiting behaviour of honey bees (*Apis mellifera* L.). Anim. Behav. **18**, 527–538 (1970)
185. Nunez, J.A.: Beobachtungen an sozialbezogenen Verhaltensweisen von Sammelbienen. Z. Tierpsychol. **28**, 1–18 (1971)
186. Nunez, J.A.: Honeybee foraging strategies at a food source in relation to its distance from the hive and the rate of sugar flow. J. Apic. Res. **21**, 139–150 (1982)
187. Nunez, J.A., Giurfa, M.: Motivation and regulation of honey bee foraging. Bee World **77**, 182–196 (1996)
188. Nürnberger, F., Keller, A., Hartel, S., Steffan-Dewenter, I.: Honey bee waggle dance communication increases diversity of pollen diets in intensively managed agricultural landscapes. Mol. Ecol. **28**, 3602–3611 (2019)

189. Nürnberger, F., Steffan-Dewenter, I., Härtel, S.: Combined effects of waggle dance communication and landscape heterogeneity on nectar and pollen uptake in honey bee colonies. Peer J. **5**, e3441 (2017)

190. Oh, S.M., Rehg, J.M., Balch, T., Dellaert, F.: Learning and inferring motion patterns using parametric segmental switching linear dynamic systems. Int. J. Comput. Vis. **77**, 103–124 (2008)

191. Okada, R., Ikeno, H., Aonuma, H., Ito, E.: Biological insights into robotics: honeybee foraging behavior by waggle dance. Adv. Robot. **22**, 1665–1681 (2018)

192. Okada, R., Ikeno, H., Sasayama, N., Aonuma, H., Kurabayashi, D., Ito, E.: The dance of the honeybee: how do they dance to transfer the food information effectively? Acta Biol. Hung. **59**, 157–162 (2018)

193. Okada, R., Akamatsu, T., Iwata, K., Ikeno, H., Kimura, T., Ohashi, M., Aonuma, H., Ito, E.: Waggle dance effect: dancing in autumn reduces the mass loss of a honeybee colony. J. Exp. Biol. **215**, 1633–1641 (2012)

194. Okada, R., Ikeno, H., Kimura, T., Ohashi, M., Aonuma, H., Ito, E.: Error in the honeybee waggle dance improves foraging flexibility. Sci. Rep. **4**, 4175 (2014)

195. Pahl, M., Tautz, J., Zhang, S.: Honeybee cognition. In: Kappeler, P. (ed.) Behavior: Evolution and Mechanisms. Springer, Heidelberg (2010)

196. Pahl, M., Zhu, H., Pix, W., Tautz, J., Zhang, S.: Circadian timed episodic-like memory—a bee knows what to do when, and also where. J. Exp. Biol. **210**, 3559–3567 (2007)

197. Pierce, A.L., Lewis, L.A., Schneider, S.: The use of the vibration signal and worker piping to influence queen behavior during Swarming in honey bees, *Apis mellifera*. Ethology **113**, 267–275 (2007)

198. Polakoff, L.M.: Dancing bees and the language controversy. Integr. Biol. **1**, 187–194 (1998)

199. Preece, K., Beekman, M.: Honeybee waggle dance error: adaptation or constraints? Unravelling the complex dance language of honeybees. Anim. Behav. **94**, 19–26 (2014)

200. Rau, A.: Realtime Honey Bee Waggle Dance Decoding System. TU Berlin, Masterarbeit (2014)

201. de Reaumur, R.A.S.: Memoires pour servir a l'historie des insectes. Imprimerie Royale, Paris (1740). Physicalisch-oeconomische Geschichte der Bienen, Deutsch. Adam Jonathan Felseckers Erben, Leipzig/Frankfurt (1759)

202. Renner, M.: Neue Versuche über den Zeitsinn der Honigbiene. Z. vergl. Physiol. **40**, 85–118 (1957)

203. Renner, M.: Über ein weiteres Versetzungsexperiment zur Analyse des Zeitsinnes und der Sonnenorientierung der Honigbiene. Z. vergl. Physiol. **42**, 449–483 (1959)

204. Renner, M.: Das Duftorgan der Honigbiene und die physiologische Bedeutung ihres Lockstoffes. Z. vergl. Physiol. **43**, 411–468 (1960)

205. Renner, M., Heinzeller, T.: Do trained honeybees with reliably blinded ocelli really return to the feeding site? J. Apic. Res. **18**, 225–229 (1979)

206. Ribbands, C.R.: The Behaviour and Social Life of Honeybees. Dover Publications, New York (1953)

207. Ribbands, C.R.: The scent perception of the honeybee. Proc. R. Soc. B **143**, 367–379 (1955)

208. Riley, J.R., Smith, A.D.: Design considerations for an harmonic radar to investigate the flight of insects at low altitude. Comput. Electron. Agr. **35**, 151–169 (2002)

209. Riley, J.R., Greggers, U., Smith, A.D., Reynolds, D.R., Menzel, R.: The flight paths of honeybees recruited by the waggle dance. Nature **435**, 205–207 (2005)

210. Riley, J.R., Smith, A.D., Reynolds, D.R., Edwards, A., Osborne, J.L., Williams, I.H., Carreck, N.L., Poppy, G.M.: Tracking bees with harmonic radar. Nature **379**, 29–30 (1999)

211. Ritschoff, C.C., Seeley, T.D.: The buzz-run: how honeybees signal ≫time to go≪. Anim. Behav. **75**, 189–197 (2008)

212. Robinson, G.E., Page, R.E., Jr.: Genetic determination of nectar foraging, pollen foraging, and nest-site scouting in honey bee colonies. Behav. Ecol. Sociobiol. **24**, 317–323 (1989)

213. Rohrseitz, K., Tautz, J.: Honey bee dance communication: waggle run direction coded in antennal contacts? J. Comp. Physiol. A **184**, 463–470 (1999)

214. Rossel, S., Wehner, R.: Celestial orientation in bees: the use of spectral cues. J. Comp. Physiol. A **155**, 605–613 (1984)

215. Rossel, S., Wehner, R.: The bee's e-vector compass. In: Menzel, R., Mercer, A. (eds.) Neurobiology and Behavior of Honeybees, pp. 76–93. Springer, Heidelberg (1987)

216. Sammlung Karl August Forster: Die Biene. Graphische Blätter aus fünf Jahrhunderten. Selbstverl., Küsnacht-Zürich (1975)

217. Sandeman, D., Tautz, J., Lindauer, M.: Transmission of vibration across honeycombs and its detection by bee leg receptors. J. Exp. Biol. **199**, 2585–2594 (1996)

218. Schäfer, S.: Schwerkrafteinfluss auf das Bewegungsmuster kurvenlaufender Bienen. Universität Ulm / Würzburg, Diplomarbeit (1997)

219. Schlegel, T., Visscher, P.K., Seeley, T.D.: Beeping and piping: characterization of two mechano-acoustic signals used by honey bees in swarming. Naturwissenschaften **99**, 1067–1071 (2012)

220. Schneider, S.S., Visscher, P.K., Camazine, S.: Vibration signal behavior or waggle-dancers in swarms of the honey bee, *Apis mellifera* (Hymenoptera: Apidea). Ethology **104**, 963–972 (1998)

221. Scholtyssek, C.: Zwei Versuche zur Bedeutung des Richtungssignals von Bienentänzen. Universität Freiburg, Diplomarbeit (1998)

222. Schöne, H.: Orientierung im Raum. Formen und Mechanismen der Lenkung des Verhaltens im Raum bei Tier und Mensch. Wissenschaftliche Verlagsgesellschaft, Stuttgart (1983)

223. Schricker, B.: Die Orientierung der Honigbiene in der Dämmerung. Zugleich ein Beitrag zur Frage der Ocellenfunktion bei Bienen. Z. vergl. Physiol. **49**, 420–458 (1965)

224. Schurch, R., Couvillon, M.J.: Too much noise on the dance floor: intra- and inter-dance angular error in honey bee waggle dances. Commun. Integr. Biol. **6**, 1–3 (2013)

225. Schurch, R., Ratnieks, F.L.W.: The spatial information content of the honey bee waggle dance. Front. Ecol. Evol. **18** (2015)

226. Schurch, R., Couvillon, M.J., Burns, D., Tasman, K., Waxman, D., Ratnieks, F.L.W.: Incorporating variability in honey bee waggle dance decoding improves the mapping of communicated resource locations. J. Comp. Physiol. A **199**, 1143–1152 (2013)

227. Seeley, T.D.: Division of labor between scouts and recruits in Honeybee foraging. Behav. Ecol. Sociobiol. **12**, 253–259 (1983)

228. Seeley, T.D.: Social foraging by honey bees: how colonies allocate foragers among patches of flowers. Behav. Ecol. Sociobiol. **19**, 343–354 (1986)

229. Seeley, T.D.: Honey bee foragers as sensory units of their colonies. Behav. Ecol. Sociobiol. **34**, 51–62 (1994)

230. Seeley, T.D.: The Wisdom of the Hive. Harvard University Press, Cambridge (1995)

231. Seeley, T.D.: Honeybee Democracy. Princeton University Press (2010)

232. Seeley, T.D., Tautz, J.: Worker piping in honey bee swarms and its role in preparing for liftoff. J. Comp. Physiol. A **187**, 667–676 (2001)

233. Seeley, T.D., Visscher, P.K.: Sensory coding of nest-site value in honeybee swarms. J. Exp. Biol. **211**, 3691–3697 (2008)

234. Seeley T.D., Camazine, S., Sneyd, J.: Collective decision-making in honey bees: how colonies choose among nectar sources. Behav. Ecol. Sociobiol. **28**, 277–290 (1991)

235. Seeley, T.D., Mikheyev, A., Pagano, G.: Dancing bees tune both duration and rate of waggle-run production in relation to nectar-source profitability. J. Comp. Physiol. A **186**, 813–819 (2000)

236. Seeley, T.D., Morse, R.A., Visscher, P.K.: The natural history of flight of honey bee swarms. Psyche **86**, 103–113 (1979)

237. Seeley, T.D., Kleinhenz, M., Bujok, B., Tautz, J.: Thorough warmup before take-off in honey bee swarms. Naturwissenschaften **90**, 256–260 (2003)

238. Sen Sarma, M., Esch, H., Tautz, J.: A comparison of the dance language in *Apis mellifera carnica* and *Apis florea* reveals striking similarities. J. Comp. Physiol. A **190**, 49–53 (2003)

239. Shannon, C.E.: A mathematical theory of communication. Bell Syst. Tech. J. **27**, 379–423 (1948)

240. Sherman, G., Visscher, P.K.: Honeybee colonies achieve fitness through dancing. Nature **419**, 920–922 (2002)

241. Simpson, J.: The mechanism of honey-bee queen piping. Z. vergl. Physiol. **48**, 277–282 (1964)

242. Sladen, F.W.L.: A scent organ in the bee. Br. Bee J. **29**, 142, 143, 151–153 (1901)

243. Srinivasan, M.V., Zhang, S., Altwein, M., Tautz, J.: Honeybee navigation: nature and calibration of the ≫odometer≪. Science **287**, 851–853 (2000)

244. Sprengel, C.K.: Das entdeckte Geheimnis der Natur im Bau und in der Befruchtung der Blumen. Vieweg, Berlin (1793)

245. Spitzner, J.E.: Ausführliche Beschreibung der Korbbienenzucht im sächsischen Churkreise. Junius, Leipzig (1788)

246. Steche, W.: Gibt es ≫Dialekte≪ der Bienensprache? Dissertation: Uni München, 1954.

247. Steche, W.: Gelenkter Bienenflug durch ≫Attrappentänze≪. Naturwissenschaften **44**, 598 (1957)

248. Steche, W.: Beiträge zur Analyse der Bienentänze. Insect. Soc. **4**, 305–318 (1957)

249. Steffan-Dewenter, I., Kuhn, A.: Honeybee foraging in differentially structured landscapes. Proc. Biol. Sci. **270**, 569–575 (2003)

250. Tanner, D.A.: An evaluation of the tuned-error hypothesis in the honey bee. MS thesis, University of California, Riverside (2003)

251. Tanner, D.A., Visscher, K.: Do honey bees tune error in their dances in nectar-foraging and house-hunting? Behav. Ecol. Sociobiol. **59**, 571–576 (2006)

252. Tanner, D.A., Visscher, P.K.: Do honey bees average directions in the waggle dance to determine a flight direction? Behav. Ecol. Sociobiol. **62**, 1891–1898 (2008)

253. Tanner, D., Visscher, K.: Does the body orientation of waggle dance followers affect the accuracy of recruitment? Apidologie **40**, 55–62 (2009)

254. Tanner, D., Visscher, P.K.: Adaptation or constraints? Reference-dependent scatter in honey bee dances. Behav. Ecol. Sociobiol. **64**, 1081–1086 (2010)

255. Tautz, J.: Honeybee waggle dance: recruitment success depends on the dance floor. J. Exp. Biol. **199**, 1375–1381 (1996)

256. Tautz, J.: Das Festnetz der Bienen. Spektrum der Wissenschaft **8**, 60–66 (2002)

257. Tautz, J.: Die Erforschung der Bienenwelt. Neue Daten – neues Wissen. Klett und Audi-Stiftung, Ingolstadt (2014)

258. Tautz, J., Rohrseitz, K.: What attracts honeybees to a waggle dancer? J. Comp. Physiol. A **183**, 661–667 (1998)

259. Tautz, J., Sandeman, D.C.: Recruitment of honeybees to non-scented food sources. J. Comp. Physiol. A **189**, 293–300 (2002)

260. Tautz, J., Casas, J., Sandeman, D.C.: Phase reversal of vibratory signals in honeycomb may assist dancing honeybees to attract their audience. J. Exp. Biol. **204**, 3737–3746 (2001)

261. Tautz, J., Rohrseitz, K., Sandeman, D.C.: One-strided waggle dance in bees. Nature **382**, 32 (1996)

262. Tautz, J., Zhang, S., Spaethe, J., Brockmann, A., Si, A., Srinivasan, M.: Honeybee odometry: performance in varying natural terrain. PLOS **2**, 0915–0923 (2004)

263. Thom, C., Gilley, D.C., Hooper, J., Esch, H.E.: The scent of the waggle dance. PLOS Biol. **5**, e228 (2007)

264. de Torres, L.M.: Tratado breve de la cultivacion y cura de las colmenas. Alcala de Henares (1586)

265. Towne, W.F.: The spatial precision and mechanisms of the dance communication of honey bees: experimental and comparative studies. Ph.D. Thesis, Princeton University Press (1985)

266. Towne, W.F., Gould, J.L.: Magnetic field sensitivity in honey bees. In: Kirschvink, J.L., Jones, D.S., McFadden, B.J. (eds.) Magnetite Biomineralization and Magenetoreception in Organisms, Chapter 18 (1984)

267. Towne, W.F., Gould, J.L.: The special precision of the honeybee's dance communication. J. Insect Behav. **1**, 129–155 (1988)

268. Towne, W.F., Kirchner, W.H.: Hearing in honey bees: detection of air-particle oscillations. Science **244**, 686–688 (1989)

269. Tsujiuchi, S., Sivan-Loukianova, E., Eberl, D.F., Kitagawa, Y., Kadowaki, T.: Dynamic range compression in the honey bee auditory system toward waggle dance sounds. PLOS ONE **2**, e234 (2007)

270. Unhoch, N.: Anleitung zur wahren Kenntnis und zweckmäßigsten Behandlung der Bienen nach drey- und dreysigjähriger genauer Beobachtung und Erfahrung. Fleischmann, München (1823)

271. Vadas, R.L.: The anatomy of an ecological controversy: honeybee searching behavior. Oikos (Forum Section) **69**, 158–166 (1994)

272. Waddington, K.D.: Honey bee foraging profitability and round dance correlates. J. Comp. Physiol. A **148**, 279–301 (1982)

273. Waddington, K.D., Kirchner, W.H.: Acoustical and behavioral correlates of profitability of food sources in honey bee round dances. Ethology **92**, 1–6 (1992)

274. Waddington, K.D., Nelson, C.M., Page, R.E., Jr.: Effects of pollen quality and genotype on the dance of foraging honey bees. Anim. Behav. **56**, 35–39 (1998)

275. Wario, F., Wild, B., Couvillon, M.J., Rojas, R., Landgraf, T.: Automatic methods for long-term tracking and the detection and decoding of communication dances in honeybees. Front. Ecol. Evol. **3** (2015)

276. Wario, F., Wild, B., Rojas, R., Landgraf, T.: Automatic detection and decoding of honey bee waggle dances. PLOS ONE **12**, e0188626 (2017)

277. Watson, J.B., Lashley, K.S.: Homing and related activities in birds. Carnegie Institution, Washington (1915)

278. Wehner, R.: Himmelsnavigation bei Insekten. Neurophysiologie und Verhalten. Neujahrsbl. Naturforsch. Ges. Zürich **184**, 1–132 (1982)

279. Wehner, R., Rossel, S.: The bee's celestial compass: a case study in behavioural neurobiology. In: Hölldobler, B., Lindauer, M. (eds.) Experimental Behavioural Ecology. Fort. Zool. **31**, 11–53. Fischer, Stuttgart/New York

280. Wehner, R., Stulzer, W., Obrist, M.: The bee-dance controversy revisited. Experientia **41**, 1223 (1985)

281. Weidenmüller, A., Seeley, T.D.: Imprecision in the waggle dances of the honeybee (*Apis mellifera*) for nearby food sources: error or adaptation? Behav. Ecol. Sociobiol. **46**, 190–199 (1999)

282. Wenner, A.M.: Sound production during the waggle dance of the honey bee. Anim. Behav. **10**, 79–95 (1962)

283. Wenner, A.M.: The elusive honey bee dance ≫language≪ hypothesis. J. Insect Behav. **15**, 859–878 (2002)

284. Wenner, A.M., Wells, P.H.: Anatomy of a controversy—the question of a ≫language≪ among bees. Columbia Press, New York (1990)

285. Wenner, A.M., Wells, P.H., Johnson, D.L.: Honeybee recruitment to food sources: olfaction or language? Science **164**, 84–86 (1969)

286. Wenner, A.M., Wells, P.H., Rohlf, F.J.: An analysis of the waggle dance and recruitment in honey bees. Physiol. Zool. **40**, 317–344 (1967)

287. Wilson, E.O.: The Insect Societies. Belknap Press of Harvard University Press, Cambridge, Mass. (1971)

288. Wray, M.K., Klein, B.A., Mattila, H.R., Seeley, T.D.: Honeybees do not reject dances for ›implausible‹ locations: reconsidering the evidence for cognitive maps in insects. Anim. Behav. **76**, 261–269 (2008)

289. Yesh'kov, Y.K., Sapozhnikov, A.M.: Mechanisms of generation and perception of electric fields by honey bees. Biofizika **21**, 1097–1102 (1976)

290. Zhang, S., Schwarz, S., Pahl, M., Zhu, H., Tautz, J.: Honeybee memory: a honey bee knows what to do and when. J. Exp. Biol. **209**, 4420–4428 (2006)

291. Zoubareff, A.: A propos d'un organe de l'abeille non encore decrit. Bull. d'Apiculture pour la Suisse Romande **5**, 215–216 (1883)